Automating Design in Pro/ ENGINEER® with Pro/ PROGRAM™

THE PROFESSIONAL USER'S GUIDE TO PROGRAMMING WITH PRO/PROGRAM

Mark Henault, Sean Sevrence, and Mike Walraven
with Fred Karam and Greg Mlotkowski

Automating Design in Pro/ ENGINEER® with Pro/ PROGRAM™

by Mark Henault, Sean Sevrence, and Mike Walraven
with Fred Karam and Greg Mlotkowski

Published by:

OnWord Press
2530 Camino Entrada
Santa Fe, NM 87505-4835 USA

First Edition, 1997

SAN 694-0269

10 9 8 7 6 5 4 3 2 1

Printed in the United States of America

Library of Congress Cataloging-in-Publication Data

Henault, Mark, 1968-
   Automating design in Pro/ENGINEER with Pro/Program /
Mark Henault, Sean Sevrence, Mike Walraven.
   p. cm.
   Includes index.
   ISBN 1-56690-117-0
   1. Pro/ENGINEER. 2. Engineering design--Data processing. 3. Computer-aided design. 4. Mechanical drawing--Data processing.
I. Sevrence, Sean, 1969- . II. Walraven, Mike, 1968-
.
III. Title.
TA174.H457   1996
620'.0042'028553--dc20   96-35767
                                                CIP

Trademarks

OnWord Press is a trademark of High Mountain Press, Inc. Pro/ENGINEER is a registered trademark of Parametric Technology Corporation. Pro/PROGRAM, Pro/ASSEMBLY, and Pro/SURFACE are trademarks of Parametric Technology Corporation. Microsoft Word is a registered trademark of Microsoft Corporation. Many other products and services are mentioned in this book that are either trademarks or registered trademarks of their respective companies. OnWord Press and the authors make no claim to these marks.

Warning and Disclaimer

This book is designed to provide information about Pro/ENGINEER and Pro/PROGRAM. Every effort has been made to make the book as complete and accurate as possible; however, no warranty or fitness is implied.

The information is provided on an "as is" basis. The authors, Parametric Technology Corporation, and OnWord Press shall have neither liability nor responsibility to any person or entity with respect to any loss or damages in connection with or rising from the information contained in this book.

The fictitious company, "Blown Away Fan Company," is just that. Any likeness to a present or former company is purely coincidental. The box fan and the round fan are for demonstration purposes only. They are not intended to be complete designs ready for manufacturing.

About the Authors

All authors have used Pro/ENGINEER for at least five years. Mike, Mark, Sean, and Greg use Pro/ENGINEER in their work for the Automotive Components Division at Ford Motor Company. Fred is an independent Pro/ENGINEER consultant who has founded his own company, Engineering Solid Solutions, Inc.

Mark Henault is a product development engineer at Ford Motor Company, where he uses Pro/ENGINEER to automate interior packaging and ergonomic studies for automobiles. He earned a B.S. and an M.S. degree in mechanical engineering from the Massachusetts Institute of Technology and is currently completing his M.B.A. at the University of Michigan. Mark previously worked for the Charles Stark Draper Laboratory as a mechanical design engineer.

Mike Walraven is a product development engineer at Ford Motor Company. He uses Pro/ENGINEER to automate the packaging of instrument panels for automobiles. Mike earned a B.S. degree in mechanical engineering from the GMI Engineering & Management Institute. Mike previously worked for Vickers as a mechanical engineer.

Sean Sevrence is a product engineering designer and uses Pro/ENGINEER to automate the packaging of instrument panels for automobiles at Ford Motor Company. He earned an A.S. degree in drafting and design from the University of Toledo in 1994.

Greg Mlotkowski is a product design engineer. He earned a B.S. in mechanical engineering from Lawrence Technological University. Greg uses Pro/ENGINEER to design center consoles and other automotive components at Ford Motor Company.

Fred A. Karam is the president of Engineering Solid Solutions, Inc. Fred holds a B.S. in mechanical engineering from Lawrence Technological University and is completing an M.S. in manufacturing systems engineering at the University of Michigan. He provides Pro/ENGINEER consulting to companies such as John Deere, Eaton, Bosch, and Ford. Fred has been a design engineer for Chrysler Corporation and a consultant for Parametric Technology Corporation.

Many Thanks

Thanks to my wife, Kerry, for her patience, understanding, and encouragement while I worked on this book. Thanks also to my family and friends who added their support and encouragement to Kerry's.

Mark Henault

Many thanks to the friends and family who made my contributions to *Automating Design in Pro/ENGINEER* possible. Special thanks to Heather for giving me the understanding and patience the book required. Last, but not least, thanks as well to my co-authors for their support and encouragement.

Mike Walraven

I would like to express many thanks to my family and friends, especially and with all my heart, to my wife Sherri, daughter Samantha, and future son, for the time to devote to this effort and the knowledge to draw from while doing so.

Sean Sevrence

Thanks to all my friends for the times I need them most.

Greg Mlotkowski

Thanks to all the guys who wrote this book for allowing me to contribute my knowledge of Pro/ENGINEER. Special thanks to my wife Chris for supporting me through a second book.

Fred Karam

OnWord Press...

OnWord Press is dedicated to the fine art of professional documentation. In addition to the authors who developed the material for this book, other members of the OnWord Press team contributed their skills to make the book a reality. Thanks to the following people and other members of the OnWord Press team who contributed to the production and distribution of this book.

Dan Raker, President
Gary Lange, Vice President
Janet Leigh Dick, Associate Publisher; Director, Channel Marketing
Daniel Clavio, Director, Market Development and Strategic Relations
David Talbott, Director of Acquisitions
Dave Surette, Senior Manager, Administration and Finance
Rena Rully, Senior Manager, Production and Editorial
Barbara Kohl, Senior Editor; Manager Book Editorial
Daril Bentley, Senior Editor
Carol Leyba, Senior Production Manager
Michelle Mann and Cynthia Welch, Production Editors
Lauri Hogan, Marketing Services Manager
Kristie Reilly, Assistant Editor
Lynne Egensteiner, Cover designer, Illustrator

Target Audience

This book is aimed at mid- to high-level Pro/ENGINEER users who want to automate the design process. Generally speaking, even the most straightforward industrial design is less than user-friendly, and few part and assembly designs, if any, are adaptable for later or different use. Furthermore, when adaption or change to parts and assemblies is desired, usually only the builders of the original geometry can execute either. The premise of this book is that adding some "high-level" logic to everyday industrial design can simplify some of this complexity. In particular, using Pro/PROGRAM can open the door to design adaptability by enabling any user familiar with Pro/ENGINEER to make geometry changes.

Inexperienced users are encouraged to read *INSIDE Pro/ENGINEER* (2nd edition, OnWord Press, 1995) and *Pro/ENGINEER Tips and Techniques* (OnWord Press, 1996) before tackling the Pro/PROGRAM procedures in this book. These complementary texts provide geometry creation tips and techniques that are especially useful for understanding and mastering the concepts contained in *Automating Pro/ENGINEER with Pro/PROGRAM*. All three books provide powerful keys to using Pro/ENGINEER for better product building.

To demonstrate how to unleash Pro/PROGRAM's power, *Automating Pro/ENGINEER* focuses on advanced tips and techniques for geometry creation. Pro/PROGRAM is a tool for programmers and non-programmers alike. Based on straightforward logic, it controls geometry construction using simple equations and controlling relations written in English. Pro/PROGRAM requires no working knowledge of C programming. It is not as powerful as Pro/Develop. Pro/Develop, however, must be compiled uniquely for different machines. Pro/PROGRAM operates on all makes of workstations.

At the time this book was being written, Parametric Technology Corporation was converting the Pro/ENGINEER user interface to dialog boxes. For this reason, *Automating Pro/ENGINEER* specifies commands but generally does not list the individual steps required to arrive at them. The overall focus is on the *functionality* used to create the parts/assemblies/programs, not the exact series of menu selections.

Typographical Conventions

Menu names appear in full capitals. Examples are the MAIN and MODIFY menus.

Menu command options appear in initial capitals such as Change Symbol and Add Member.

User input, and names for files, directories, tables, variables, parameters, parts, assemblies, and so on in regular text are italicized. Examples follow:

- Rename the length dimension of the pipe from *d1* to *pipe_length*.
- If the overall length of the generic *motor_screw.prt* changes, the instances retain their values specified in the table.

- This feature can be seen by retrieving either the *motor_1* or *motor_2* instance.

Pro/PROGRAM code lines are shown in a monospaced typeface as follows:

```
input
passed_value number
  "what is the value that you wish to pass to the parts?"
    end input
```

✓ **TIP:** *Tips on functionality usage, shortcuts, and other information aimed at saving you time and toil appear like this.*

➼ **NOTE:** *Information on features and tasks that is not immediately obvious or intuitive appears in notes.*

☞ **WARNING:** *Warnings appear in the book to help prevent you from committing yourself to unexpected or undesired outcomes.*

Companion CD-ROM

The companion CD-ROM contains the Pro/ENGINEER and Microsoft Word files linked to the examples throughout the text. Use the files for hands-on demonstrations of Pro/PROGRAM's ability to solve identical geometry problems in a multitude of ways. Some solutions are more graceful than others, but each reflects Pro/PROGRAM's flexibility.

The uncompressed files on the CD total approximately 33 Mb. The Pro/ENGINEER files are Version 16 and will also run in Version 17. The Microsoft Word files are Version 6.0 for Windows and will run only on IBM-compatible PCs. If you use a Macintosh computer you will need to use either conversion templates or a DOS emulator. Other files are in ASCII format and will operate on any hardware. If loading the files causes problems, try following the commands for loading files to specific machines as described in your hardware manual or as prescribed by a systems administrator.

File Descriptions

Top level assembly

fan.asm	Top level assembly controlling both the box fan assembly (inexpensive fan) and the round fan assembly (mid-range fan).	248 Kb

Subassemblies

alt_blade_motor.asm	Modified blade_motor.asm used in Chapter 8.	28 Kb
family_prog_ex.asm	Assembly used in family tables chapter.	17 Kb
inter_prog_ex.asm	Assembly used in the interchange groups chapter.	13 Kb
sw_buttons_base_mold_asm.asm	Assembly for mold creation in Chapter 10.	7435 Kb
sw_buttons_base_mold_asm_.mfg	Manufacturing file for mold creation in Chapter 10.	2 Kb
alt_box_fan.asm	Alternate method of creating a Pro/PROGRAM using interchange groups and family table parts. Used in Chapter 8 to show another example of how to organize a Pro/program.	214 Kb

sw_dial.asm	Rotary switch subassembly used for both types of fans.	27 Kb
sw_buttons.asm	Pushbutton switch subassembly used for both types of fans.	29 Kb
sw_rocker.asm	Rocker switch subassembly used for both types of fans.	28 Kb
switches.asm	Interchange group used in Chapter 8. Controls the three unique switch types in the top level assembly.	16 Kb
blade_motor.asm	Subassembly for the blade and motor chosen before being assembled into the top level fan assembly.	28 Kb

Box fan parts (quantity used)

box_fan.prt	Steel housing for the box fans (1).	633 Kb
box_fan_grill.prt	Plastic grill parts that prevent fingers from coming in contact with the fan blade. Used in conjunction with the box fan (2).	1815 Kb
box_fan_foam.prt	Shipping foam used for the box fan assemblies (1).	137 Kb
box_fan_box.prt	Shipping box used when creating a box fan assembly (1).	56 Kb
foot.prt	Stabilizing foot used for the box fans (2).	144 Kb

box_fan_screws.prt	Screws to connect the box_grill to the box_fan part (8).	76 Kb
box_fan_vert_members.prt	Vertical steel member welded to the box fan housing that the motor bolts to for support (2).	153 Kb
alt_blades.prt	Modified blades.prt used in Chapter 8.	823 Kb
alt_box_fan_housing.prt	Modified box_fan_housing.prt used in Chapter 8.	665 Kb
alt_box_fan_vert_members.prt	Modified box_fan_vert_members.prt used in Chapter 8.	150 Kb
alt_motor.prt	Modified motor.prt used in Chapter 8.	360 Kb

Round fan parts (quantity used)

front_plastic_housing.prt	Front injection molded housing for the round fan. Basically functions as a finger guard (1).	2340 Kb
rear_plastic_housing.prt	Rear injection molded housing for the round fan. Basically functions as a finger guard (1).	5139 Kb
base.prt	Base, or one-piece leg part, that holds the round fan assembly up (1).	1096 Kb
round_fan_foam.prt	Shipping foam used for the round fan assemblies (1).	437 Kb
round_fan_box.prt	Shipping box used when creating a round fan assembly (1).	64 Kb

Common parts for both fans (quantity used)

switch_screw.prt	Screws that attach the switches to the fan housings (2 to 4).	71 Kb
motor_screw.prt	Screws that attach the motors to the fans (4).	81 Kb
motor.prt	Family table part for the five motors used by Blown Away Fan Company for both box and round floor models (1).	354 Kb
blades.prt	Fan blades used for both box and round floor fans. In this case, even though Blown Away Fan Company only has five blade choices, this part is fully functional with a separate Pro/program. For purposes of this book, the part will assume five unique configurations although it is far more flexible (1).	652 Kb
sw_buttons.prt	Buttons in the sw_buttons assembly (1).	1005 Kb
sw_buttons_hsg_base.prt	Button switch housing base (1).	764 Kb
sw_buttons_hsg_cover.prt	Button switch housing cover (1).	327 Kb
sw_dial.prt	Rotary switch (1).	1145 Kb
sw_dial_hsg.prt	Rotary switch housing (1).	428 Kb
sw_rocker.prt	Rocker switch (1).	706 Kb
sw_rocker_hsg.prt	Rocker switch housing (1).	323 Kb
sw_rocker_pin.prt	Rocker switch pivot axis pin (1).	36 Kb

Assembly used for drawings

fan1.asm	Specific version of the box fan subassembly previously mentioned. Provided to allow user to call up the drawing described in the text and linked initially to the Microsoft Word file.	174 Kb

Miscellaneous Parts

actual_molding.prt	Untrimmed part released by the mold, including all runners and molded parts.	913 Kb
sw_button_hsg_fea.prt	Version of the button housing part that contains the mesh controls used in the finite element analysis section of Chapter 10.	321 Kb
sw_buttons_hsg_base_ref.prt	First reference copy of the sw_buttons_hsg_base.prt created during mold creation in Chapter 10 (5% shrinkage factor).	629 Kb
sw_buttons_hsg_base_ref2.prt	Second reference copy of the sw_buttons_hsg_base.prt created during mold creation in Chapter 10 (15% shrinkage factor).	629 Kb
overhead.prt	Bulk part created to store overhead values used in cost.drw.	5 Kb

mold_workpiece.prt	Solid part created to represent entire mold in Chapter 10.	42 Kb
mold_top.prt	Top half of mold extracted in Chapter 10.	578 Kb
mold_bottom.prt	Bottom half of mold extracted in Chapter 10.	290 Kb
labor_cost.prt	Bulk part created to store labor costs used in cost.drw.	5 Kb

Drawings

fan.drw	Multi-page drawing linked to the process sheets contained in the fan2.doc Word document. Shows the proper format for linking to Microsoft Word.	95 Kb
cost.drw	Drawing file containing the table which estimates total cost of fan assembly.	113 Kb
motor_screw.drw	Tabular drawing of family table parts shown in first illustration of Chapter 3.	35 Kb

Microsoft Word Files

fan1.doc	Linked Microsoft Word document used in Chapter 9. This file is linked to the fan.drw file above.	61 Kb
fan2.doc	Contains process sheets discussed in Chapter 9. This file is linked to the Pro/ENGINEER drawing, fan.drw.	480 Kb

Table of Contents

Chapter 3:
Pro/PROGRAM and Family Tables 35

Chapter 4:
Interchange Groups 57

Chapter 5:
Assembly Modeling 73

Chapter 6:
Advanced Pro/PROGRAM Functionality 93

Chapter 7:
Pulling It All Together Without
Family Tables and Interchange Groups 107

Introduction

Why Pro/PROGRAM?

Using Pro/ENGINEER to automate the design process of common parts and Pro/PROGRAM to create customized input menus is an extremely powerful way to leverage productivity. A few moments spent in Pro/ENGINEER and Pro/PROGRAM can save hours in work load redundancy. For example, creating an interchange assembly or capturing design intent in a Pro/program gives the keys for modifying complex parts to users who may not know how to create them. Pro/PROGRAM is a useful tool for training and for model modification. It creates straightforward menus that give users access to modifiable dimensions and enables modification without requiring precise knowledge of the design intent or geometry parameters. Time usually spent on training or sending the object back to its original builder can be applied to other, more productive tasks.

Pro/programs are particularly useful for businesses that want to productively use the expertise of diverse employee groups or divisions. For example, a company comprised of two divisions, one making fan switches and another fan housings, would most likely have one group extremely knowledgeable in the manufacture of housings and another knowledgeable in the making of switches. Can one division modify the designs of the other as needed? Can the people in housing adapt switches to meet housing specifications and vice versa? By capturing design intent and controlling the actual design changes,

Pro/programs make these types of cross-divisional changes possible.

Numerous methods can combine available components in different ways to create multiple assembly variations. One Pro/ENGINEER user might manually create each assembly and modify the geometry as needed for different versions. Another user might make interchange groups and/or family tables of components that someone experienced in Pro/ENGINEER could quickly alter. However, each of these approaches depends on knowledge of how to correctly retrieve and assemble components.

In contrast, Pro/PROGRAM users actually *automate* parts assembly, reducing modification to a series of a few selections. Pro/PROGRAM users create Pro/programs that manage features and components to rebuild geometry so that even new users can make complex changes. With Pro/PROGRAM fewer employees can study more design iterations in less time because of the inherent design knowledge captured in the program.

The main questions to ask when building a part or assembly in Pro/ENGINEER follow:

❖ Will this component be used again?

❖ Will changes be required by persons other than the creator of the part?

❖ Will automating the design allow less capable users to make major changes to the part/assembly and thereby become more productive?

If the answer to any of the above questions is yes, then setting up a user-friendly menu and description of the critical design parameters using Pro/PRO-GRAM probably makes sense. Pro/PROGRAM users can create parts and assembly models that new or inexperienced users can modify without first having to dig through the entire design to find the "correct" feature to change. For example, a Pro/program could document the dimensions of a "box" in such a way that new users would immediately know the parameters' meanings, and the implications of changing them. Appearing below is a sample description of dimensions in Pro/PROGRAM.

```
VERSION 16.0

REVNUM 665

LISTING FOR PART BOX


INPUT

width number

"what is the width of the box?"

height number

"what is the height of the box?"

depth number

"what is the depth of the box?"

END INPUT
```

Pro/PROGRAM and Pro/ENGINEER

Pro/PROGRAM operates via numerical inputs or requests for yes/no answers. Once the numbers are keyed (the queries are answered), the program automatically recreates the geometry or performs the component assembly or disassembly. Users can also enter input strings to select preexisting Pro/ENGINEER parts or assemblies by name. The comment line provides users with a more detailed description of the requested input. (See Chapter 2 for details.)

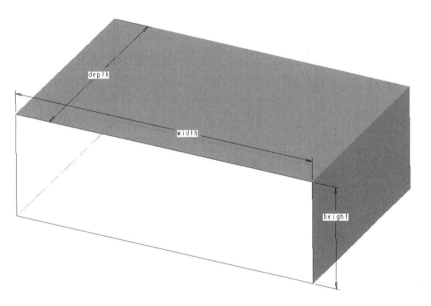

A simple box showing dimensions.

Pro/ENGINEER offers many ways to enhance an organization's power to quickly evaluate multiple iterations of assemblies or part designs. The focus throughout *Automating Pro/ENGINEER* is on demonstrating the many ways in which Pro/PROGRAM enables organizations to harness this power by automating the design process.

Automating Pro/ENGINEER walks you through creating Pro/programs to generate and control mid- to high-level Pro/ENGINEER parts and assemblies. The book uses several comprehensive examples (available on the companion CD) based on the fictitious business organization, Blown Away Fan Company (BAF).

Pro/PROGRAM and Automation

BAF manufactures two very different fan lines: an inexpensive line of box fans—with many different combinations of switches, motors, and blade configurations—and a not-quite-as-inexpensive line of round floor fans with a separate support base. BAF purchases motors, fan blades, and switches from suppliers. The company manufactures the rest of the components required to make and ship its floor fans.

BAF wants to offer the most cost-effective fan it can for sale next summer. Consequently, the company would like to create and evaluate multiple floor fan prototypes using a top level assembly in Pro/ENGINEER.

BAF also publishes its own technical documentation using PCs and a variety of software packages. Published documents include an assembly manual for in-house use by line workers, consumer warranty information, and final assembly instructions for consumers.

What to Automate

Box Fan Components

- ❖ Housing (steel)

- ❖ Blades (choice of 5)

- ❖ Motor (choice of 5)

- ❖ Vertical stiffeners to which the motor mounts (2)

- ❖ Finger guard grills (2)

- ❖ Tip prevention feet (2)

- ❖ Switches (choice of 3)

- ❖ Grill screws (8)

- ❖ Motor screws (4)

- ❖ Switch screws (2–4)

- ❖ Packing foam

- ❖ Shipping box

Box fan assembly.

Round Fan Components

❖ Two-piece case (plastic; finger guard grills molded in)

❖ Blades (choice of 5)

❖ Motor (choice of 5)

❖ Tip prevention base

❖ Switches (choice of 3)

❖ Motor screws (4)

❖ Switch screws (2 to 4)

❖ Packing foam

❖ Shipping box

Round fan assembly.

Various techniques are employed throughout the book to create and control required geometry: family tables, interchange groups, relations, and Pro/PRO-GRAM. Unique Pro/programs control each part, and a single top level assembly links the individual parts and programs and controls the available variations.

The fictional assembly created for this book automatically generates over 90 unique fans, along with their shipping foam and boxes. Pro/PROGRAM also reports the estimated cost to manufacture and ship the product, and Pro/ENGINEER files linked to PC applications enable rapid updating of geometry files within instruction manuals.

Automation Step by Step

Step 1

Model the limited number of available motors using family tables. This step effectively *stores* all conceivable variations for retrieval as needed.

"Typical" motor part.

Step 2

Use relations to control the fan housing geometry and the fan blade. Set up the relations such that users with general knowledge of manufacturing rules can alter the parts with ease.

Box fan housing.

"Typical" fan blade.

Step 3

Create and use interchange groups to set up the control of switches in the top level assembly. Consider potential pitfalls in interchange group use.

Three switch assemblies in interchange group.

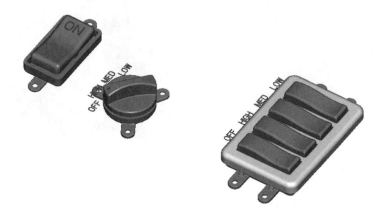

Step 4

Explore advanced geometry techniques in the creation of round fan housings and box fan grills. Create a program to control the geometry creation phase. Explore assembly modeling techniques that greatly enhance design flexibility.

Round fan housing front.

Box fan grill.

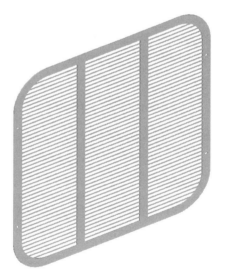

Step 5

Put Pro/PROGRAM to the test: automatically rebuild
the geometry of the switches, motors, and housings
created in steps 1 through 4. Create a Pro/program
that prompts the user for required inputs and then
makes changes to geometry without the user having
to know which feature to select. An example Pro/
PROGRAM input menu appears below.

```
VERSION 16.0

REVNUM 10886

LISTING FOR GENERIC PART BOX_FAN_GRILL


INPUT

FAN_DIA NUMBER

"WHAT IS THE FAN BLADE DIAMETER?"

GRILL_DEPTH NUMBER

"ENTER THE DEPTH OF THE GRILL"

GRILL_FLANGE_WIDTH NUMBER

"ENTER THE WIDTH OF THE PERIMETER FLANGE FOR THE GRILL (RECO 25)"

RIB_HEIGHT NUMBER

"ENTER THE THICKNESS OF THE RIBS (RECO 2.5)"

VERTICAL_STIFFENERS NUMBER

"ENTER THE NUMBER OF VERTICAL STIFFENERS WANTED (RECO 2)"

CORNER_RADII NUMBER

"WHAT ARE THE CORNER RADII?"

END INPUT
```

Step 6

Greatly enhance the Pro/program to incorporate fan blades, shipping foam, and shipping box into the top level assembly. Use Pro/PROGRAM to estimate the shipping and manufacturing costs associated with each fan variation created. Quickly iterate fan prototypes and their costs. Select the most cost-effective fan for meeting customer needs.

Round fan shipping foam.

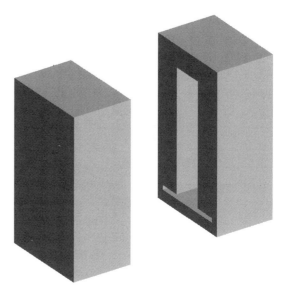

Step 7

Link Pro/ENGINEER drawings to PC applications to enable rapid documentation change corresponding to changes in geometry. Incorporate Pro/ENGINEER data into Microsoft Word, CorelDRAW, Designer, WordPerfect, and Visual REAL applications.

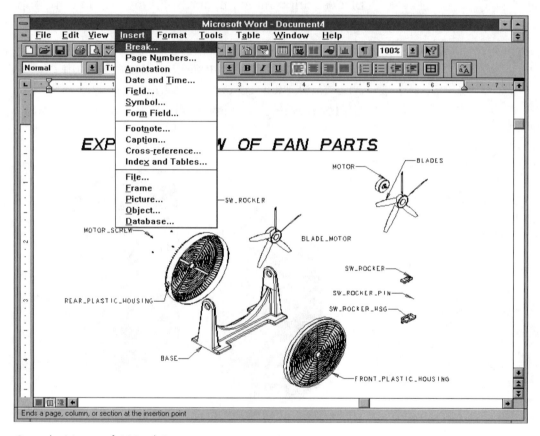

Sample Microsoft Word Screen.

Step 8

Finish automating the design by creating parametric molds. Use Pro/FEM to create a parametric mesh for analysis and create *.stl* files for rapid prototyping. All files will be linked to the original design such that any changes driven by the Pro/program will automatically cascade throughout the entire design.

Parametric mold.

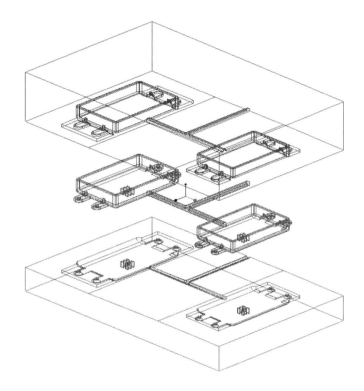

Summary

This chapter presented a quick overview of book contents. The book has been organized (with all data files provided on the companion CD-ROM) in such a way that you can quickly and effectively learn how to use the power of Pro/PROGRAM to automate the design process. By the time you complete the book, we are confident that Pro/PROGRAM will become part of your design toolkit.

Pro/PROGRAM
Basics and
Sample File

This chapter covers Pro/PROGRAM basics and the concepts underlying the creation of relations and the use of Pro/PROGRAM to automate them. Subsequent chapters examine more advanced topics.

Pro/PROGRAM captures the design intent of design or assembly models by automating their "engineering intelligence." In some respects Pro/PROGRAM acts as an automated family table, but without a table's complex and cumbersome design matrices.

Pro/programs achieve different design iterations by selecting and changing design variables. For example, a Pro/program for a CD player might automate the size, shape, and cosmetic symbols assigned to buttons. The program might also prompt installation of a small or large disk carrier.

Pro/ programming

Many users become discouraged upon hearing the word "program" linked to automated processes. Pro/PROGRAM uses a simple BASIC–like programming interface to create design variations, but it is not an application programmer's interface (API). The user requires no programming knowledge. Pro/PROGRAM controls the variables assigned to each relation within a feature or component by linking them interactively within a simple program.

Defining and executing a Pro/program creates a model that responds automatically to design specification changes. Pro/PROGRAM'S user interface requests input information and passes it directly into the design for regeneration. By simply executing a Pro/program and responding to its prompts for information, a novice user can make design changes. The same user can also instantly review each iteration for feedback on changes in cost, weight, bill of material, and package size.

Relations

Relations are numerical equations that link dimensions or parameters within a model to help capture the desired design intent. Pro/PROGRAM automates this process by creating user-defined programs that control and drive relations and design parameters. Parametric relations are used to control the design relationships of a part or an assembly.

Pro/PROGRAM users modify feature relations numerically or conditionally. Relations are defined by equality or by comparison.

❖ *Equality.* Equates a design variable to a fixed numerical value. For example, an equality relation would set the *pipe_length* parameter as equal to 30 mm (*pipe_length* = 30).

❖ *Comparison.* Assigns value using If...Else...Endif statements. An example follows:

```
IF pipe_dia = 3
pipe_length = 5*pipe_dia
ENDIF
```

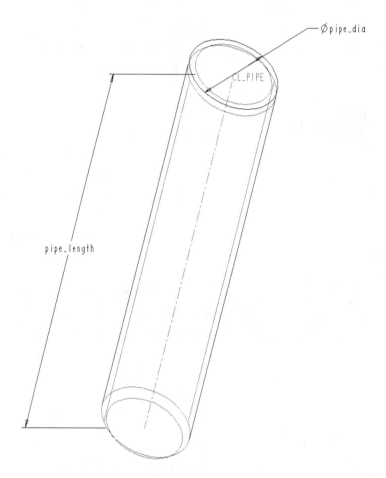

Pipe parameters.

Design Variables

Design variables are a model's "identifiers," that is, the parameters algebraic equations use to control how features, parts, or assemblies react to one another. A Pro/program running on a model changes the model's design variables. Changing the variables changes the topology (shape) of the design.

Input Variables

Input variables are used to identify and track key inputs from within the program model. Pro/PRO-GRAM supports three types of input variables. These types can also act as user-defined parameters.

❖ **Number.** Identifies a numerical input from the keyboard.

❖ **String.** Identifies a string of characters that allows input of parameters or model name.

❖ **Yes_No.** Identifies either a YES or NO input statement for the defined variable type.

↝ *NOTE: The name and type of the variable must be entered between INPUT and END INPUT.*

User-defined parameters are ideal for declaring additional information within a model. For example, if color is a significant factor in a design, a parameter may be defined (e.g., Color = Red) to identify the color of a model. Such parameters are not necessarily involved in a model's relation.

Editing and Initiating Pro/ programs

Pro/program editing and initiating occur directly from within the part or assembly model. Select Edit Design and then From Model or From File.

❖ **From Model** instructions and features correspond to the current state of the model.

❖ **From File** retrieves the last edited version of the program file stored as an existing *.als or *.pls file. These files are created when you edit a program, but do not incorporate the changes made in the model file.

The following prompt appears after a Pro/program is constructed: "incorporate the design changes into the model Yes or No." A No response creates a file called *modelname.als* or *modelname.pls*. This file appears as an additional menu the next time Edit Design is selected.

Prompts

System prompts are used as a way to interface with the program. These prompts are displayed in the information window located at the bottom of the display. A prompt requests information from the user for use in the model. User-defined prompts are created and customized within the Pro/PROGRAM environment.

↝ **NOTE:** *Prompts must immediately follow the input statement and must be contained within quotation marks.*

Pro/ENGINEER Dimensions

When Pro/ENGINEER creates a model it sequentially assigns a default internal symbolic ID number such as d0, d1, or d2 to each dimension.

✓ *TIP: Renaming the symbolic ID to correspond with the Pro/PROGRAM input variable avoids having to create a relation that ties the two together because Pro/ENGINEER automatically passes the value through the model. Rename using the Change Symbol command located under the DIM COSMETICS menu.*

Pro/ENGINEER supports the following dimension types.

❖ **Part (d#).** Dimensions assigned within the part that are created from the sketcher relations once the feature has been regenerated.

❖ **Assembly (d# ; #).** Part feature dimensions displayed in the assembly environment. The trailing number corresponds to the part assembly sequence order.

❖ **Reference (rd#).** Model dimensions that do not drive geometry but provide numerical feedback as the design changes. For instance, the reference *overall_length* could specify a model's overall length for use in other relations.

❖ **Sketcher (sd#).** Dimensions assigned in the sketcher mode for use with known dimensions to build control over a sketched feature.

❖ **Known (kd#).** Dimensions assigned within the sketcher mode to predefined, regenerated geometry. Known dimensions are not attached to sketched entities and are not factored in solving the sketched cross section. Users can relate sketched entities to known dimensions to make them responsive to the design intent.

➥ *NOTE: Sketcher dimensions have two significant drawbacks for use with Pro/PROGRAM. (1) If a feature includes sketcher relations, the only ways to access its relations are to redefine its sketched entities or to modify the relations via Relations | Feature | Sketcher. (2) Features with sketched entities take extra regeneration time during Pro/ PROGRAM execution.*

✓ *TIP: Toggle the symbolic ID and the numeric value of the dimension back and forth by selecting the Switch Dim command under the RELATIONS menu.*

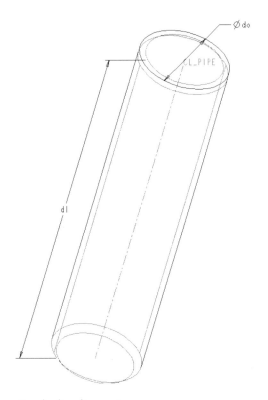

Symbolic dimensions.

Model Information

Five basic areas of Pro/PROGRAM model information appear in the program editor used for Pro/PRO-
GRAM development.

Area	Information
Header	Version, revision number, and listing for part or assembly
Input	Parameters that attach information in the model to the executable program
Relations	Basic relations as well as relations that tie the input parameters to the symbolic ID dimensions
Model Listing	Controls that add features or parts based on a conditional statement
Mass Property	Definitions and updates for mass properties related to geometry changes

Sample Program: Pipe Part

This sample program controls the length of a pipe based on the input parameters for the pipe's diameter. The program also prompts whether to include chamfers on the ends of the pipe.

Pipe part.

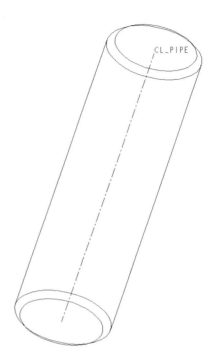

1. Create a pipe of diameter 3 mm and length of 20 mm by extruding a circular cross section.

2. Create a chamfer of 0.25 mm on the pipe ends.

3. Rename the diametrical dimension symbol ID from *d0* to *pipe_dia* using the Change Symbol command located under the DIM COSMETICS menu.

4. Rename the length dimension of the pipe from *d1* to *pipe_length*.

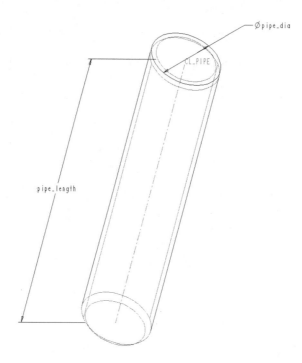

Symbolic values changing.

5. Initiate a Pro/program From Model and modify the file at the input block, relations block, and chamfer feature as follows:

```
VERSION 16.0

REVNUM 45

LISTING FOR PART PIPE


INPUT

PIPE_DIA NUMBER

"WHAT IS THE DIAMETER OF THE PIPE?"

CHAMFER YES_NO
```

"DO YOU WANT A CHAMFER AT BOTH ENDS?"

END INPUT

RELATIONS

PIPE_LENGTH=3*PIPE_DIA

END RELATIONS

ADD FEATURE (initial number 1)

INTERNAL FEATURE ID 1

FIRST FEATURE: Extrude

NO.	ELEMENT NAME	INFO	STATUS
1	Section	Defined	
2	Depth	Blind, depth = 9	Defined

SECTION NAME = S2D0002

FEATURE'S DIMENSIONS:

pipe_dia = 3.00 Dia

pipe_length = 9.00

END ADD

IF CHAMFER == YES

ADD FEATURE

INTERNAL FEATURE ID 14

```
PARENTS = 1(#1)


CHAMFER: Edge Chamfer


NO.     ELEMENT NAME    INFO            STATUS
1       Scheme          45 x d          Defined
2       Edge Refs       2 curve or edge(s)  Defined


FEATURE'S DIMENSIONS:
d2 = .25
END ADD
END IF


MASSPROP
END MASSPROP
```

6. Save the file within the editor and answer "yes" when asked if you wish to incorporate the changes into the model. You will then be queried as to whether you wish to use Current Values or Enter new values. Choose Enter and a new window will appear showing the two inputs.

7. Select both the *Pipe_dia* and *Chamfer* inputs, pick Done, and press Enter. Pro/ENGINEER will then ask you for new values for the two inputs (one at a time).

Parameter window.

8. Enter the desired shaft diameter and whether the pipe ends are to have chamfers. Pro/ENGINEER passes the values to the model, evaluates the relations, adds or suppresses the chamfers, and regenerates the part.

➼ ***NOTE:*** *As the mouse moves over an input parameter, the current value displays in yellow at the bottom of the information window along with the program prompt.*

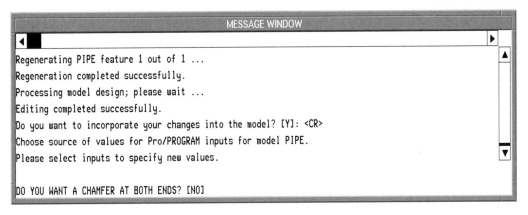

Information window.

➼ ***NOTE:*** *Identifying and renaming the parameters within the model before writing the program* *reduces the number of relations required in the Pro/program.*

Summary

This chapter covered the basic structure of a Pro/program. The next several chapters introduce components controlled by a Pro/program. You are encouraged to retrieve the model files from the companion CD-ROM and examine the program as you proceed through the book. Chapter 6 introduces advanced Pro/PROGRAM concepts.

Pro/PROGRAM
and Family Tables

Family tables are very powerful but underused Pro/ENGINEER tools that have a variety of applications. This chapter describes selected major applications in detail and touches briefly on others. The concept behind all of them is similar, but the goals for each are slightly different.

Determining when to use a family table can be tricky. Asking the following questions should help in decision making:

❖ Do you have "common" standard parts such as bolts or nuts that have similar geometry, but are different in size?

❖ Do you wish for design modifications to take place by changing one generic part and automatically updating 100-plus instances rather than 100-plus parts individually?

❖ Do you have user-defined features such as screw bosses that you would like to make available to your design team, and if changes to the "standard" boss design are necessary, all people using them would automatically get updated?

❖ Do you have several similar designs that you would like to be able to seamlessly interchange within an assembly?

❖ Would you like the capability of creating separate drawings for individual parts in a "family of parts" quickly and easily?

If you answered yes to any of the above questions, family tables are probably for you.

Managing Parts and Assemblies

In essence, family tables are an organizing concept used to manage features, dimensions, parameters, and so on, within "families" of parts and assemblies. The following sections focus on selected applications and benefits of family tables for generic parts and instances, common "standard parts," and common generic features.

Generic Parts and Instances

Generic parts contain a component's basic design information such as required features, relations, parameters, and so on common to all family-table-driven parts or instances. Benefits of the generic part/instance relationship follow:

❖ Information or detail common to many similar parts need only be entered once.

❖ Automatic interchange within assemblies and drawings.

❖ Fewer part files because only the generic model is stored.

❖ Easy tabular drawing creation.

Non-repetition of Common Information

Consider, for example, a standard bolt. A parameter calling out the material is added to the generic part, and all instances automatically contain the part. Without a family table, the parameter would have to be entered for each individual bolt part created. For hundreds of components driven from one generic part, this capability offers serious time savings.

Automatic Interchange

Because the IDs for features in the generic part and the instance are identical, they can be automatically interchanged within assemblies and drawings. This feature can save enormous amounts of time when creating individual drawings for instances. In effect, a single generic drawing can be made and copied to a new name, the model can be replaced with one of the instances, and all dimensions, parameters, and so on will be automatically updated. This process completes virtually all the work required to create the drawing, and only a small amount of cleanup is necessary.

Automatic interchange within assemblies is also powerful. For example, if both exist in the same family table, a short bolt in an assembly can be interchanged automatically with a longer bolt. All assembly constraints are maintained, and any other components relying on that bolt for location or geometry creation purposes are automatically reassembled or regenerated. The only other methods for creating this interchange are layout (notebook) and interchange groups, both of which require more work to set up than warranted by a small number of parts.

Fewer Part Files

One benefit of the generic-to-instance relationship is that all information required to build the instance resides within the generic model and its family table. Consequently, Pro/ENGINEER does not need to store each instance as a separate part file. Instead, the program stores only the generic model and creates an index file for the instances, and lists their names and corresponding generic part names. This file allows instance retrieval directly by name without pulling up the generic part first. The file is simply a pointer file within the directory where the generic part is stored, and can be deleted.

➥ **NOTE:** *If index files are deleted, the only way to retrieve an instance whose generic part is not in session is to retrieve the generic part first.*

Because Pro/ENGINEER must first retrieve the generic part, copy it to the instance, and then regenerate the instance to its family table values, instance retrieval times can be lengthy. To reduce retrieval time, PTC added an option in Release 16 that stores the instances to disk. Thus, retrieval no longer requires regeneration of the instance. The option consists of instance accelerator files (stored instances) which require much more disk space. For example, a part with 50 instances would require approximately 50 times the disk space.

✓ **TIP:** *For simple parts (e.g., less than 100 features) the authors recommend not using instance accelerator files. The reduction in generation time simply is not significant enough to warrant the extra disk space required.*

GENERIC PART NAME: MOTOR_SCREW	
	MODEL PARAMETERS
INSTANCE NAME	LENGTH
MOTOR_SCREW_1	14.000
MOTOR_SCREW_2	16.000
MOTOR_SCREW_3	18.000
MOTOR_SCREW_4	20.000
MOTOR_SCREW_5	22.000

SCALE : 3:1   TYPE : PART   NAME : MOTOR_SCREW SIZE : A

Tabular drawing of family table parts.

Simple Tabular Drawing Creation

Because the family table contains all dimension values for the instances, drawings can display values and parameters associated with every instance in an accompanying table. This alternative to having individual drawings for instances should benefit anyone not absolutely requiring individual drawings. Changes made to the family table are automatically updated on the drawing, and the table creation is semi-automated using a feature called Repeat Regions.

Common "Standard Parts"

Common "standard parts" are truly a "family of parts," with common geometry yet different sizes. These types of designs lend themselves extremely well to family tables. Many companies have tabulated drawings of such components. A family table is the way to make the geometry available within Pro/ENGINEER. Component dimensions are about the only things entered in the family table.

The *motor_screw.prt* file on the companion CD-ROM is a generic part example. The overall length of the screw is the only dimension entered in the family table. (This is, of course, a very simple example chosen only to show functionality within family tables.)

motor_screw.prt and overall length dimension.

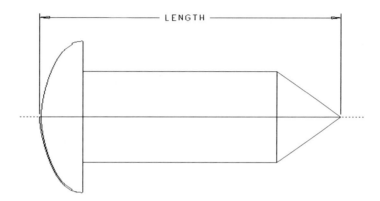

The family table for the generic part example shows five instances named along the left column as *motor_screw_1* to *motor_screw_5*. The five individual parts are controlled from the generic part.

```
┌──────────────────────────────────────────────────────────────────────────────┐
│ ─   Pro/TABLE TM Release 16.0 (c) 1988-95 by Parametric Technology Corporation All Rights Reserved. │
├──────────────────────────────────────────────────────────────────────────────┤
│ File  Edit  View  Format  Help                                                 │
├──────────────────────────────────────────────────────────────────────────────┤
│ !                                                                              │
├──────────────────────────────────────────────────────────────────────────────┤
│        C1 C2         C3        C4        C5        C6        C7       C8      C9 │
│ R14  ! 9) You may add more entries to the bottom of the table as needed.       │
│ R15  ! 10) Pro/TABLE formatting characters will also be ignored.               │
│ R16  ! 11) Feature identifications are their internal ids.                     │
│ R17  !                                                                         │
│ R18  !   Generic part name: MOTOR_SCREW                                        │
│ R19  ! Name                d16                                                 │
│ R20  !                     LENGTH                                              │
│ R21  ! ============== ========== ==========                                    │
│ R22  ! GENERIC        14.0000                                                  │
│ R23    MOTOR_SCREW_1   14.0000                                                 │
│ R24    MOTOR_SCREW_2   16.0000                                                 │
│ R25    MOTOR_SCREW_3   18.0000                                                 │
│ R26    MOTOR_SCREW_4   20.0000                                                 │
│ R27    MOTOR_SCREW_5   22.0000                                                 │
│ R28                                                                            │
│ R29                                                                            │
│ R30                                                                            │
│ R31                                                                            │
│ R32                                                                            │
│ R33                                                                            │
│ R34                                                                            │
│ R35                                                                            │
│ R36                                                                            │
│ R37                                                                            │
│ R38                                                                            │
│ R39                                                                            │
└──────────────────────────────────────────────────────────────────────────────┘
```

Family table of motor_screw.prt.

The second column displays the overall length dimension associated with each screw. All five parts will exhibit any changes made to the generic part, except for changes to dimensions driven by the family table. For example, if the size of the screw head were to increase, all five instances would also change to reflect the same. However, dimensions controlled in the family table do not change when the generic part changes. If the overall length of the generic *motor_screw.prt* changes, the instances retain the values specified in the table.

↝ **NOTE 1:** *Instance dimensions contained within the family table may be changed in two ways. First, you can modify dimensions while editing the family table. Second, while in the instance, select MODIFY and answer Yes to the prompt to confirm that you want to change the dimension value in the table. Modification of the table-driven dimension values for the generic model can be executed only while in the generic model. Modification of non-table-driven dimension values can be executed in either the generic part or the instance, and will be updated in both.*

from parent

↝ **NOTE 2:** *Dimensions are not the only data that you enter in family tables. Parameters, tolerances, text strings, and so on may also be entered.*

✓ **TIP:** *To clarify the dimension to be changed while editing, select the MODIFY menu and choose Dim Cosmetics | Symbol to rename the dimension value added to the table. In the example, d16 was changed to LENGTH. Note that the original dimension symbol and the new, modified symbol both exist in the table.*

Common "Generic Features"

Generic features can be defined a number of ways. For our purposes, they are defined as follows:

❖ Features common to several similar components

❖ UDFs (user-defined features)

Features Common to Several Similar Components

To examine this definition in greater detail, turn again to the generic part *motor.prt*.

Generic motor.prt.

One of the most difficult (but powerful) concepts to grasp when using family tables is the capability to contain instances that have features other instances do not. This capability allows for the possibility of a base generic part with instances having very different shapes and features.

A simple example appears in the Pro/ENGINEER manual (*Pro/ENGINEER Fundamentals,* Parametric Technology Corp., 1989–1995). The example, multi-level family tables for screws, shows that the basic generic part contains only the screw shaft. The next level down contains a fillister head or a round head added to two different instances. These features are totally different and could not be present at the same time in the generic part because they would overlap.

Nonetheless, both instances are developed from the same generic part. Therefore, a round-headed screw and a fillister-headed screw have the same advantages as all other family-table-driven parts. They can be assembled using a feature common to both, and are therefore readily interchangeable. A generic drawing setup can be used to create individual drawings for both types of screws, with the detailing required only for the features unique to that type of instance (round or fillister).

Now back to the *motor.prt.* Note that the generic motor has no electrical connection. The generic motor lacks a connector because the two motor sizes require different connectors result of horsepower (HP) differentials. Because both connectors are located on the motor's top side, a regeneration error would occur if you were to add both features to model the connectors.

Generic motor without electrical connection.

Two sets of features were created as the workaround. One set is for the smaller HP motors and consists of an external protrusion with three pins and several rounds that simulate the female portion of a connector. This feature can be seen by retrieving either the *motor_1* or *motor_2* instance.

Motor_1 instance showing electrical connector.

The second set of features contains a cut into the outer diameter of the motor, 10 pins, and some rounds. This feature can be seen by the retrieving *motor_3*, *motor_4*, or *motor_5* instance.

Motor_3 instance showing electrical connector.

The above two connectors clearly have completely different geometry, and if both sets of features in the generic part were turned on, a regeneration error would probably occur. Therefore, both sets of features are suppressed within the generic part. In the family table, the features ELECT_CONN_1 and ELECT_CONN_2 are suppressed for the generic part. ELECT CONN_1 is resumed for instances *motor_1* and *motor_2* and suppressed for *motor_3*, *motor_4*, and *motor_5*. Conversely, ELECT_CONN_2 is suppressed for instances *motor_1* and *motor_2*, and resumed for *motor_3*, *motor_4*, and *motor_5*.

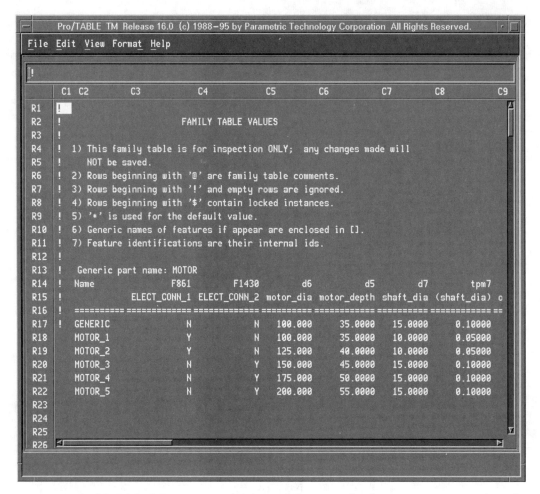

Note status of ELECT_CONN_1 and ELECT_CONN_2 for the generic part and each instance.

✓ **TIP:** *Grouping features into local groups and placing only one of a group's features in the family table will (1) dramatically reduce the number of columns in the family table, and (2) make it much easier to determine which features to include in each instance.*

It should be noted that there is more than one way to create this type of geometry. While working in the generic part, the features can be added one at a time, grouped, and added to the family table. You would work on one of the groups, put it in the family table, suppress it, and then create the other one. This is an acceptable way of adding features, but there is a better, more efficient method. It is possible to add the features directly to the family table instance. A definite benefit to this approach is that adding the features directly to the instance automatically adds them to the family table, suppresses them for the generic model and the other instances in the family table, and turns them on for the instance to which they are added.

∞ **NOTE:** *When adding new features directly to the instance, features are placed in the family table as separate columns. If you group the new features and wish for only one of them to show up in the family table, you will have to manually edit the family table to delete the extra feature columns.*

UDFs (User-defined Features)

A good example of a UDF is a screw boss used for plastic injection molded components. UDFs can also be set up using family tables, such that there is a single generic screw boss and many instances for items such as screw diameters and lengths. The base design for all possible bosses is the same, and can be handled as outlined in the "Common 'Standard Parts'" section. There is, however, one major difference in that generic UDFs cannot be placed on a component; an instance must be chosen. Next, modifying a UDF can only be accomplished by modifying the group itself, and using the Group Table command.

Family Table Techniques

The possibilities for creating and using family tables are virtually endless. However, the following techniques can save time with any creation method: Instantiate, choosing an instance using Pro/PROGRAM, and using a finite element analysis (FEA) model as a family table instance.

Instantiate

Instantiate is a Pro/PROGRAM command that can help facilitate the creation of family table instances in assemblies. Selected after regenerating the assembly and entering the Pro/PROGRAM inputs to formulate it, Instantiate automatically saves the assembly configuration into a family table. When invoked from the PROGRAM menu, the user must enter the instance names for subassemblies and parts into the table. Using the Instantiate command adds If loops around EXECUTE statements in the Pro/program for the assembly. The command also adds columns in the table asking whether or not to execute the part or subassembly.

Choosing an Instance

Due to the inherent interchangeability of instances within an assembly, in some cases it makes sense to have users choose which instance they want through the use of a Pro/program. This is easily accomplished by assigning a variable in an input statement as the instance name. Then in the ADD PART statement in the Pro/program, replace the part name with the variable name enclosed in parentheses. An example using the assembly *family_prog_ex.asm* is shown in the next illustration.

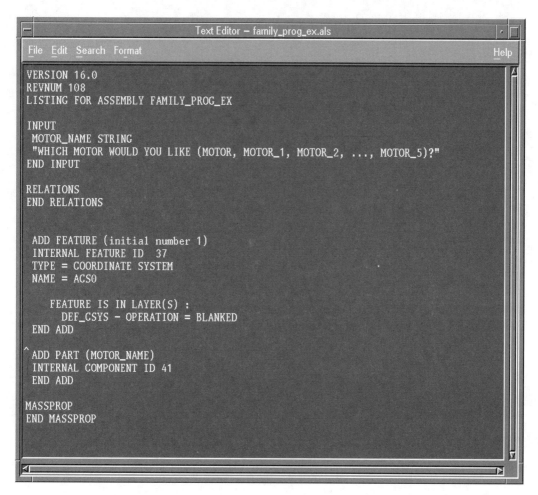

```
┌─────────────────────────────────────────────────────────────────┐
│ ─ │              Text Editor – family_prog_ex.als              │·│□│
├─────────────────────────────────────────────────────────────────┤
│ File  Edit  Search  Format                                  Help │
│ VERSION 16.0                                                      │
│ REVNUM 108                                                       │
│ LISTING FOR ASSEMBLY FAMILY_PROG_EX                              │
│                                                                  │
│ INPUT                                                            │
│  MOTOR_NAME STRING                                               │
│  "WHICH MOTOR WOULD YOU LIKE (MOTOR, MOTOR_1, MOTOR_2, ..., MOTOR_5)?" │
│ END INPUT                                                        │
│                                                                  │
│ RELATIONS                                                        │
│ END RELATIONS                                                    │
│                                                                  │
│                                                                  │
│  ADD FEATURE (initial number 1)                                 │
│  INTERNAL FEATURE ID  37                                        │
│  TYPE = COORDINATE SYSTEM                                        │
│  NAME = ACS0                                                     │
│                                                                  │
│     FEATURE IS IN LAYER(S) :                                    │
│        DEF_CSYS – OPERATION = BLANKED                           │
│  END ADD                                                        │
│                                                                  │
│  ADD PART (MOTOR_NAME)                                          │
│  INTERNAL COMPONENT ID 41                                       │
│  END ADD                                                        │
│                                                                  │
│ MASSPROP                                                        │
│ END MASSPROP                                                    │
│                                                                  │
└─────────────────────────────────────────────────────────────────┘
```

Pro/PROGRAM for assembly family_prog_ex.asm.

In the Input section of the Pro/program is an assignment to the variable MOTOR_NAME. This string variable is in effect the part name of the motor for the assembly. The ADD PART statement has been changed from ADD PART MOTOR to use the variable name instead. The new statement ADD PART (MOTOR_NAME) tells Pro/ENGINEER to add the part with name MOTOR_NAME. This will only work for the generic part and instances. If the name of another part that is not a member of the family table is entered, Pro/ENGINEER will not know how to assemble it into the assembly and therefore will not be able to replace it.

➡ *NOTE 1:* *The functionality limiting replacement of members to other family table members is only true for multiple component assemblies. Because the assembly provided is a single component assembly, and no other components follow or reference it, the component can be exchanged for any other component. Just regenerate and input the name of the component you want. The authors know of no real use for this functionality.*

➡ *NOTE 2:* *When interchanging using Pro/PROGRAM with other objects such as interchange group components, change the ADD PART statement to ADD COMPONENT to allow parts or subassemblies to be added to the assembly.*

FEA Model As Instance

Another interesting use of family tables/instances is the possibility of converting the instance to an FEA model of the generic part. This method allows the easy creation at any time of the FEA model by simply retrieving the instance. The features that you wish to suppress in order to create the FEA model would simply be placed in a family table. You can then create an "FEA instance" for which these features are set to "NO" in the family table. When the FEA instance is retrieved, the features would be suppressed.

Summary

Family tables are very powerful tools with an endless variety of uses. In this chapter family tables and reasons for their use were defined. Next, selected uses and methods of creating family tables were discussed.

4

Interchange
Groups

This chapter explains how to use Pro/ENGINEER's Interchange mode (available in the Pro/ASSEMBLY module) to change or "swap" parts and assemblies, including those with different geometries. The examples demonstrate setting up interchange groups and geometry "tagging," Pro/ENGINEER's identification process for automatic parts switching.

Interchange group switches.asm.

Using Interchange Groups

Interchange makes some design tasks easy and quick, and makes other tasks possible. With the use of Interchange, a designer can, for instance, view different wheels on the same car model, check how a new railing changes a deck, or simplify an assembly using representative shapes.

Interchange groups are collections of parts that can be interchanged for rapid design revision. A major difference between interchange groups and standard assemblies is that interchange groups declare all parts equal, whether similar or dissimilar. In other words, a cylinder that would be two parts in a standard assembly can be a square in an interchange group. Interchange groups maximize Pro/ENGINEER's advantages as follows:

❖ **Quick assembly part interchange.** Swapping parts in and out of an assembly makes real-time design review possible. Multiple studies can be accomplished in the middle of a management review. No more delays to redo and replot studies; no more follow-up meetings to review them.

❖ **Assembly simplification.** Simplifying model detail reduces the time required for regeneration and display. Components with less detail also require less memory. This is another great advantage for management design reviews: no more delays while waiting for the image to generate on-screen.

✓ *TIP: Swapping out detail intensive assemblies for "detail-light" assemblies can speed up regeneration and display times required for large assemblies.*

Interchange Groups and Pro/PDM

Pro/PDM is Pro/ENGINEER's data management tool. Pro/PDM fetching retrieves data from the network disk to your local disk. Pro/PROGRAM configures the fan assembly in the demonstration module so that selecting various components produces the overall fan. When fetching an assembly from PDM, you specify the amount of data to retrieve: All, Required, or None. Selecting None retrieves only the assembly file, Required retrieves only the parts required to regenerate the assembly at its stored state, and All fetches every file the assembly references. If a part referenced by the assembly belongs to an interchange group, the interchange group and related parts are also retrieved.

You can selectively turn parts off in an assembly. Pro/PDM will retrieve such components only when they are part of an interchange group. If they are not part of an interchange group, Pro/PDM will not retrieve them even if instructed to retrieve All. Using an interchange group signals Pro/PDM to retrieve components for both the *fan.asm* and the interchange group. Although the interchange group is not assembled in the *fan.asm,* Pro/PDM fetches all required components. When creating interchange groups, keep the following in mind:

❖ **Assembly retrieval.** Interchange groups have the same extension as an assembly (*filename.asm*), but can only be retrieved in Interchange mode. Likewise, Assembly mode assemblies can only be retrieved in Assembly mode.

❖ **Dbms/Save As.** This command does not create a part that is interchangeable with the original. In other words, you cannot use Save As to create an interchange group. You must assemble the desired parts into the interchange group, and then identify the entities to be considered equal.

❖ **Restricted assembly creation.** Members of assemblies are already interchangeable by using Simplified Reps; therefore, interchange groups should be created only when needed.

Creating Interchange Groups

This chapter demonstrates how to create an interchange group by recreating the *switches.asm* interchange group included on the CD.

1. To create an interchange group, select Interchange | Create and enter a name. For this example, use the name *switch_types*.

2. A second menu appears for specifying the interchange group type. *Functional interchange* defines a set of parts and assemblies that are functionally the same. These interchange groups are changed using the Replace option in the EDIT REP menu. Reference tags must be created and assigned to the corresponding entities in order for the components to be swapped in the assembly.

Simplify interchange defines representations for parts or assemblies that contain different information such as detailed, envelope, schematic, or symbolic. These interchange groups are changed using the Simplify option in the EDIT REP menu. The user assembles, packages, and creates the components in this interchange. Mass properties can be assigned throughout each of the representations to ensure the proper calculation in the assembly regardless of which representation is active.

Select interchange type Functional.

3. Select Add Member to begin creating the interchange group. Pick the *sw_rocker.asm* as the first component.

Assembly sw_rocker.asm.

↝ **NOTE**: *This type of assembly does not require that components be constrained. However, references*

must be set after
added to the inter

_ _ _ _ _ _ _ _ _ _are

4. Using Add M
   assemblies, su.
   the interchange gi.

Assemblies sw_rocker
and sw_dial.asm.

Interchange group
switches.asm.

Display1EcouponAction.actio...

5. Define the reference tags that will allow Pro/ENGINEER to swap the components by selecting ReferenceTag | Tags | Create.

6. Enter a reference tag name. For this demonstration, use the name *SW_LOCATOR*.

7. Assign the entities of each component that will be equivalent by selecting Assign | SW_LOCATOR. Select the coordinate system in each switch assembly called *assy_sw_rocker, assy_sw_dial,* and *assy_sw_buttons*. Once these three are chosen, select DoneSel to stop adding entities to the reference tag. Select Info in the TAGS menu to review which entities are assigned to the reference tag. The information window showing the entities assigned to the SW_LOCATOR reference tag appears below.

```
INTERCHANGE NAME = SWITCHES
REFERENCE NAME = SW_LOCATOR
REFERENCE TYPE = COORDINATE SYSTEM
THE REFERENCE TAG IS ASSIGNED TO
   SW_ROCKER (ASSEMBLY)
   SW_DIAL (ASSEMBLY)
   SW_BUTTONS (PART)
```

Review

The preceding steps created the functional interchange group *switch_type.asm* by adding the required components, creating a reference tag, and setting the entities to be considered equivalent.

↝ **NOTE:** *Chapter 4 of the* Assembly Modeling User's Guide *contains detailed information on Interchange mode functionality.*

Interchange and Pro/ PROGRAM

Once an interchange group is created, incorporating it into Pro/PROGRAM fulfills its capability to swap parts. Creating a Pro/program is not necessary; however, it can make swapping parts easier, especially for a less experienced user. Appearing below is the Pro/program from *inter_prog_ex.asm*.

```
VERSION 16.0
REVNUM 38
LISTING FOR ASSEMBLY INTER_PROG_EX


INPUT
  SWITCH_TYPE NUMBER
  "ENTER SWITCH CHOICE (1=ROCKER, 2=DIAL, 3=BUTTONS)"
END INPUT


RELATIONS
IF SWITCH_TYPE==1
```

```
SWITCH_NAME = "SW_ROCKER.ASM"

ELSE

IF SWITCH_TYPE==2

SWITCH_NAME = "SW_DIAL.ASM"

ELSE

IF SWITCH_TYPE==3

SWITCH_NAME = "SW_BUTTONS.ASM"

ENDIF

ENDIF

ENDIF

END RELATIONS


  ADD FEATURE (initial number 1)

  INTERNAL FEATURE ID 12

  TYPE = COORDINATE SYSTEM

  NAME = ACSØ


    FEATURE IS IN LAYER(S) :

        6_ALL_NON_GEOMETRY_FEATURES - OPERATION = NORMAL

        6___ALL_CSYS - OPERATION = NORMAL

  END ADD


  ADD COMPONENT (SWITCH_NAME)

  INTERNAL COMPONENT ID 14

  END ADD
```

```
MASSPROP

END MASSPROP
```

Incorporating an Interchange Group

The *inter_prog_ex.asm* sample assembly shows how simply Pro/PROGRAM can incorporate an interchange group, beginning with adding the interchange group to the overall assembly.

☛ **WARNING:** *In order for the switches to understand how to assemble themselves, assembling the interchange group must be accomplished using the reference tag.*

The following steps incorporate the interchange group into the assembly.

1. In the *fan.asm* file, assemble any of the three switches.

2. In the INPUT section, add the variable name and specify that it is a number variable. Next, add the prompt query to the user as shown below.

```
INPUT
  SWITCH_TYPE NUMBER
  "ENTER SWITCH CHOICE (1=ROCKER, 2=DIAL, 3=BUTTONS)"
END INPUT
```

↝ **NOTE:** *Pro/PROGRAM automatically capitalizes the variable name and type. Users must enter* *capital letters within the quotation marks for the questions.*

3. In the RELATIONS section, add the If statements that define which component is set to the *switch_type* variable. Add the lines shown below.

```
RELATIONS
IF SWITCH_TYPE==1
SWITCH_NAME = "SW.ROCKER.ASM"
ELSE
IF SWITCH_TYPE==2
SWITCH_NAME = "SW.DIAL.ASM"
ELSE
IF SWITCH_TYPE==3
SWITCH_NAME = "SW_BUTTONS.ASM"
ENDIF
ENDIF
ENDIF
END RELATIONS
```

4. Finally, in the Pro/PROGRAM area where the component is added, change the component name to the variable name.

```
ADD COMPONENT (SWITCH_NAME)
INTERNAL COMPONENT ID 14
END ADD
```

Interchange Group Limitations

The functionality of interchange groups is great; however, a few major deficiencies are noteworthy.

Reordering Incapability

All components must be assembled before reference tags are created and assigned. This may not be problematic for the initial creation but very problematic for maintenance. A possible workaround follows:

1. Delete all reference tags.

2. Add the new components.

3. Recreate the reference tags.

4. Reassign the reference tags.

But even the workaround has the potential for causing a lot of work and headaches, especially, for example, if an assembly of 50 requires a single new screw.

☛ *WARNING: All components must be assembled before reference tags are created and assigned; otherwise, circular references will be created.*

Extensive Memory Requirements

The entire interchange group must be in memory to swap components. The computer must have enough RAM and swap space available to work with a top level assembly and all interchange groups to swap parts.

Geometry Cannot Reference Interchange Components

Simply stated, no features can be built referencing components of an interchange assembly. Well, you could do so; however, the feature would fail once the component was swapped out of the assembly.

↔ **NOTE:** *This functionality was never intended to work. An enhancement request has been submitted to PTC to make the functionality operable. It was not a component of the application when shipped.*

This lack of functionality was discovered while attempting to attach the wiring to the switches in the *alt_fan.asm*. It was hoped that a datum point could be placed on each switch and the wire attached to the datum point. The functionality in question would make the interchange group more powerful.

The datum point example can be made effective by hardcoding the values for the point into Pro/PRO-GRAM. To accomplish this, take the following steps:

1. Place a datum point on each switch where the wiring would exit the housing.

2. In the overall assembly, swap in each switch and record the incremental distance between a coordinate system and each datum point created in step 1.

3. Create an assembly feature datum point using Offset Csys and entering one set of values.

4. Edit the Pro/Program. After the switch selected statement, insert a line to set the values for the assembly datum point to the values you recorded in step 2.

Because the assembly datum point is only being relocated, the wiring can now be set up to reference the datum point. Chapter 5 describes assembly features in detail.

Summary

In this chapter interchange groups were used to swap assemblies, even in cases of completely different geometry. This was accomplished by "tagging" geometry in the separate components, a process that informs Pro/ENGINEER on how to reassemble them. The final topic was incorporating the interchange group into a Pro/program.

Assembly
Modeling

Assembly level features allow rapid changes to occur without creating relations or programs to control them. Created while viewing an assembly and modifying a part within it, assembly features reference the geometry of other parts. As long as the geometry (or internal ID) remains intact, the assembly features will automatically reconnect themselves when the reference parts move or change.

Choose when to use assembly features carefully. Pro/ENGINEER's associativity is defined such that every feature—datum point, curve, axis, or surface—has an internal identification number. When you use a feature belonging to another part, that part's ID is stored in the feature creation. If you can guarantee that the reference feature you select will not be eliminated (e.g., through a machining operation or a fillet that eliminates a surface), you can use the feature to help achieve automatic updates to parts.

Controlling Curves: Part Level

Pro/ENGINEER encompasses a multitude of assembly features. This chapter illustrates their power by concentrating on assembly features for datum curves—a feature type common to many models.

Datum curves are key to robust modeling. Because curve definition is mathematically simple, the software handles it with ease. Using a curve edge to create other features (surfaces, protrusions, cuts, and so on) creates a security blanket in the rare event that a feature fails. Using the curve also ensures that other features or parts will use the same logic. This chapter explores the following types of controlling curves:

❖ Part level curves that other parts reference

❖ Assembly level curves that control other parts (also known as *skeleton curves*)

1. All three switch assemblies use part level curves. Access the *sw_rocker_hsg.prt* file and turn on the *sw_curve* layer.

sw_rocker_hsg shown with datum curve.

2. The datum curve was the first feature in the part, with the exception of the datum planes and coord-system. This curve is the size of the actual switch lever. The housing, therefore, must provide clearance to the outside of the curve to allow the switch to rotate. To offset the housing from

the datum curve, select Geom Tools | Offset
Edge on the curve while sketching the protrusion
in the housing.

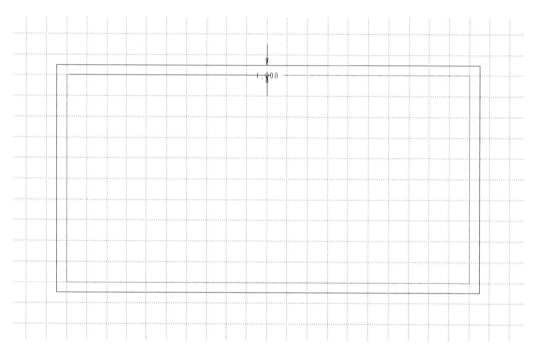

Sketch of the housing protrusion.

3. Fillet the protrusion, and then shell it to the outside so as to maintain the clearance to the datum
curve.

NOTE: *Shelling the part after the fillets have been
made automatically creates the fillets on the
other side of the material thickness. This saves
time and is good modeling practice. Controlling
the fillets with just a few dimensions makes*

updates easier to maintain, especially when the material thickness is changed.

4. Model the mounting ears and put the final rounds in place. Once *sw_rocker_hsg.prt* is finished, the *sw_rocker.asm* is created. To create the assembly, select Assembly | Create from the MODE menu.

5. Create the default datum planes and default coordinate system; then assemble the *sw_rocker_hsg* using the coordinate system in each. The assembly should resemble the next illustration.

View of assembly after adding the rocker switch housing.

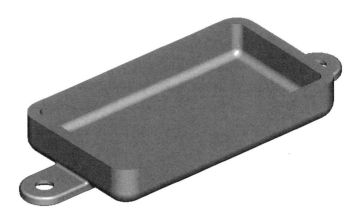

6. Create a *sw_rocker_pin* part. Create the default datum planes and default coordinate system, and then assemble it into the *sw_rocker* assembly using the coordinate systems in each.

7. Begin modeling the pin around which the rocker switch will rotate by selecting MODIFY | Mod Part | Sel By Menu | SW_ROCKER_PIN. You are now modifying the *sw_rocker_pin* part while viewing the *sw_rocker* assembly.

☛ **WARNING:** *While modifying a part in an assembly, it is very important to select the datum planes from the part being modified for the Sketcher. Not doing so could impact functionality, resulting, for example, in parts that update properly only from the assembly. Such dysfunctionality may seem insignificant, but it can create problems, particularly if someone tries to modify such parts from outside the assembly.*

8. Create the pin protrusion. Sketch on the *sw_rocker_pin_x* and use the *sw_rocker_pin_y* to the top. Change into the sketch view and sketch a circle such that the center is on the *sw_rocker_pin_z* plane and inside the *sw_rocker_hsg* part, as shown in the following illustration.

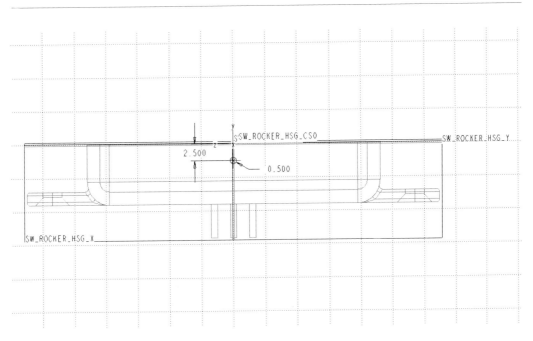

Sketch of pin protrusion.

9. Dimension the pin 2.5 mm below the *sw_rocker_ pin_y* plane and a 0.5-mm radius. Regenerate the sketch and change into the default view.

10. Select Done. For the depth, select Up To Surface. Select the outside back surface of the housing.

Outside back surface of
sw_rocker_hsg part.

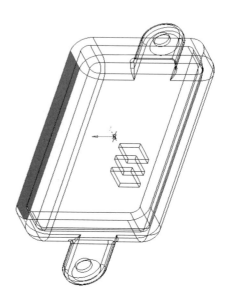

11. Select Up To Surface again and pick the outside front surface of the housing. Once the protrusion is created, you will see that the the pin extends to each outside surface of the *swrocker_hsg* part.

Outside front surface of sw_rocker_hsg part.

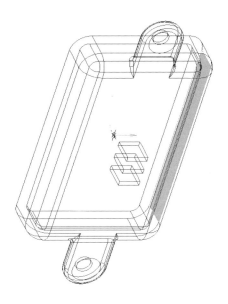

sw_rocker_assembly with the new sw_rocker_pin.

12. Modify the *sw_rocker_hsg* to add a hole to
accommodate the pin. While still in the assembly,
select Modify | Mod Part, and choose the *sw_
rocker_hsg* part. Next, create a cut sketched on
the *sw_rocker_hsg_x* plane with the *sw_rocker_
hsg_y* to the top. Use Geom Tools | Offset Edge
on the end of the *sw_rocker_pin* and a value of
.005 for the offset.

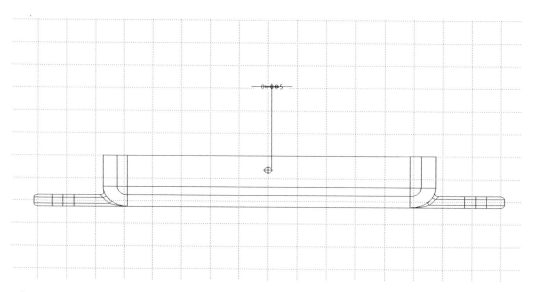

Sketch of clearance hole in sw_rocker_hsg part.

Congratulations, you have just created two
assembly level features.

⊸ **NOTE:** *Select references from other parts—known
as external references—only when needed such
as in setting the depth of the pin. It is best to have
as few external references as possible because the*

*assembly must be active and set to Regenerate |
Automatic for the depth of the pin to update.
With no or few external references, another user
making the design changes will not need the tem-
porary assembly you created, but only the part
requiring change.*

Controlling Curves: Assembly Level

Assembly level curves, as the name suggests, are
assembly features of datum curves. Their functionality is the same as part level datum curves. Because
circular references have a tendency to crop up when
using part level curves, assembly level curves may be
a better choice, particularly if you intend to have
multiple parts reference the same curve. Reasons for
using assembly level curves appear below

❖ **Master section development.** Developing a section of an assembly before creating parts can
create a blueprint for what will be modeled. The
assembly datum curve can drive the shape of the
geometry in the parts that will make up the
assembly. Once the master section is developed,
the parts in the assembly will use the assembly
curve during the sketch creation of the geometry.
Therefore, when the assembly curve is modified,
the parts will also modify simultaneously.

❖ **Constraint development.** Using only a datum curve makes constraint development simple. It makes concept development—exploring how the assembly should look and work—easier as well.

❖ **Skeleton curves.** Assembly curves can also be used to control the placement of components in the assembly. This is accomplished by first sketching an assembly datum curve, then placing a coordinate system on the datum curve. The parts can then be assembled to the coordinate system. The placement of components can also be controlled by using an offset coordinate system instead of datum curves. Either way, the skeleton technique can significantly enhance assembly performance.

Designing Clearances

In the switch assemblies, there are two instances of clearance maintenance. The first, the clearance of the rocker to the housing, is covered in the "Controlling Curves: Part Level" section. The second is the clearance of the strengthening ribs on the mounting tabs in *sw_buttons.asm*.

sw_buttons.asm.

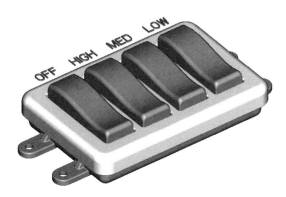

Each rib has to maintain a clearance to both the switch cover and the mounting boss. This section demonstrates how to create the rib clearance to the switch cover.

1. While viewing the *sw_buttons* assembly, create a protrusion on a plane through the middle of the mounting ear extruding to both sides.

2. Sketch a vertical centerline tangent to the outboard edge of the *sw_buttons_ hsg_cover* part. Align the centerline to that edge.

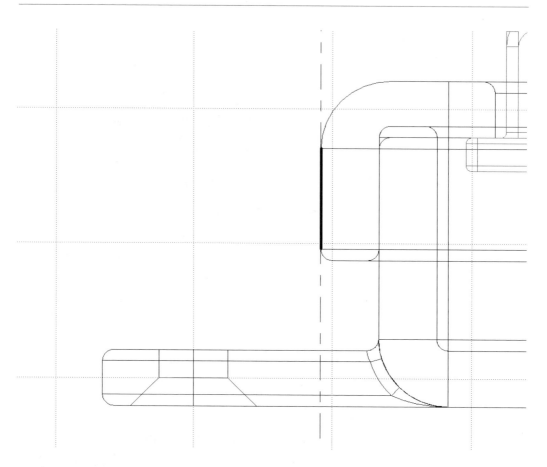

sw_buttons.asm.

3. Sketch a point where the centerline and the top of the mounting ear intersect. Align the point to the top of the mounting ear.

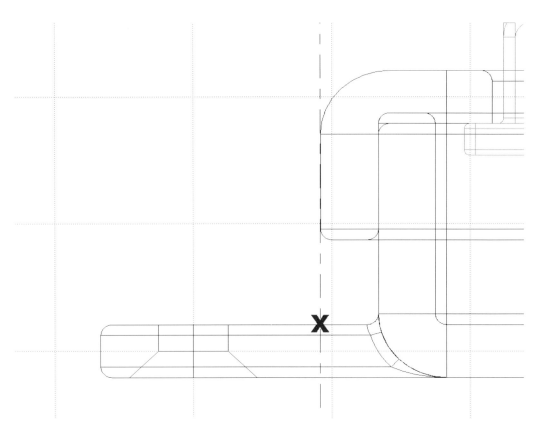

sw_buttons.asm.

4. Sketch a point on the outboard vertical wall of the *sw_buttons_hsg_base* part such that the point will be lower than the bottom edge of the *sw_buttons_hsg_cover.*

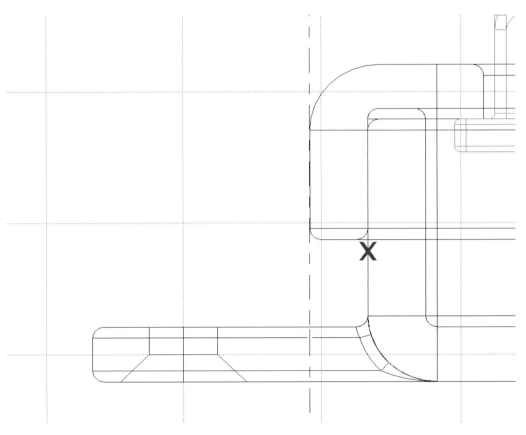

sw_buttons.asm.

5. Align the point to the vertical wall. Dimension the point to the bottom edge of the *sw_buttons_ hsg_cover.*

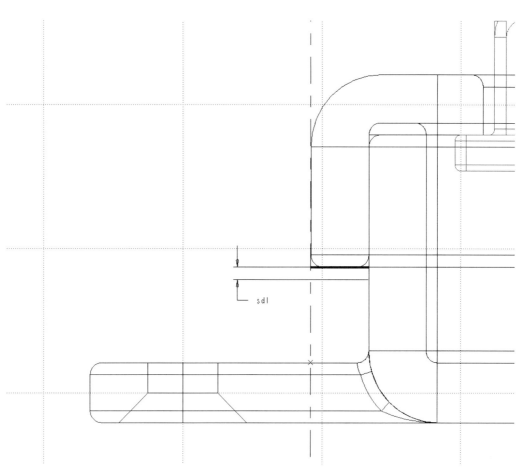

sw_buttons.asm.

6. Connect the points with a line, and close off the triangle by sketching a vertical and horizontal line.

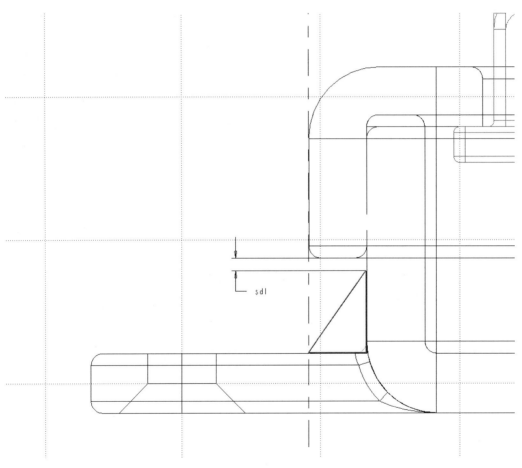

sw_buttons.asm.

7. Regenerate the sketch and modify the dimension to .5. Select Done and extrude the section to a depth of 2.

You have now created an assembly feature that the rib will update based on the design of the *sw_buttons_hsg_cover.* If the material thickness changes, the rib will follow. If the overlap of the *sw_buttons_hsg_cover* changes, the rib will follow.

The above example provides a sample of the power that Pro/ENGINEER provides with associativity. The rib is associated to the *sw_buttons_hsg_cover* because of the dimensioning technique used. To achieve a different reaction of the rib, modifying the dimensioning scheme is all that is required.

Summary

Assembly level features provide functionality equal to that of sketcher relations and part relations, without having to create any relations. The use of controlling curves in an assembly will enable ease of modifying not only the part geometry, but also the part position.

Because assembly level features directly feed part creation and all controlling elements are contained in the assembly, it is very easy for complex assembly level changes to be controlled by a Pro/program. The user can control an assembly level curve, for example, via a Pro/program which can then propagate changes down to multiple parts in the assembly

without the user having to know where to look for
the dimensions. Refer to Chapters 2 and 6 for a complete description of Pro/PROGRAM.

Advanced Pro/PROGRAM Functionality

As Chapter 2 and the examples in Chapters 3 to 5 demonstrate, Pro/PROGRAM is powerful, but that does not mean it cannot be problematic. The beginning of this chapter covers a few problems Pro/PROGRAM users may encounter. The solutions presented should make the program easier to use. Later sections of the chapter cover advanced techniques for putting Pro/PROGRAM to work in parts and assemblies.

Potential Problems

Problems can present themselves as you begin to use Pro/PROGRAM. Recreating the examples in this book may already have introduced you to a few. Avoiding possible pitfalls is one way to make learning Pro/PROGRAM easier. A good place to begin the discussion of advanced functionality is a focus on the following potential problem areas:

❖ SGI Editor (JOT) problems

❖ Naming of variables

❖ Turning parts "off" in an assembly

❖ Common errors while creating programs

SGI Editor Problems

Launched from within Pro/ENGINEER, the JOT editor can introduce a peculiar problem if you make an error while editing a Pro/program. Normally, after you invoke the Pro/program editor, create the Pro/

program, and save and exit the Pro/program, Pro/ENGINEER asks you whether you want to incorporate the changes into your part. If the Pro/program contains an error, when you exit you will be told that there is an error in the program and asked if you wish to re-edit the file. Here is the problem: If you answer "yes," no matter how many times you fix the problem, save the file, and exit again, you will continue to be told that the Pro/program contains an error and you will be asked whether you want to re-edit or abort.

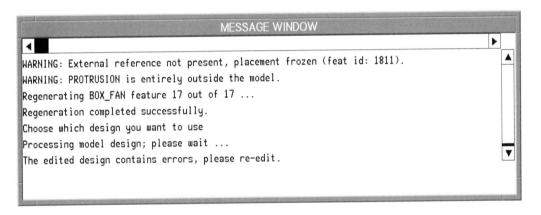

Message window showing an error in a Pro/program.

There are three ways out of this loop. The first, and not recommended, method involves aborting the file. After you choose Abort, you re-edit the Pro/program and choose "from model." A notification appears that this file, *filename.pls,* already exists and asks if you wish to overwrite it. Answering yes brings up an "original" version of the Pro/program for re-editing. However, all the work you did on the file when you created the error will be lost.

The second, and preferred, method is to save the file with the error and, when Pro/ENGINEER asks whether to re-edit or abort, choose Abort. This saves *filename.pls* in the directory where you launched Pro/ENGINEER. Then open a new system window and use the JOT editor (or any editor of your choice) to open the *.pls* file and correct problems. Save the corrected file and return to Pro/ENGINEER. Edit the program and choose From File to open the corrected version. Save and exit again and the Pro/program should execute without problems.

A third way to avoid the editing problem is to use the VI editor instead of the JOT editor.

Naming of Variables

Another frequently encountered problem arises from naming parameters during parts creation. As mentioned in Chapter 2, using descriptive names is good practice, especially for making a part easier to understand. But there are correct and incorrect ways to perform naming. The examples that follow show which is which.

Correct Method for Naming

Before editing the Pro/program, choose MODIFY | Dim Symbol and change the dimension symbol names as follows:

❖ d0 to width

❖ d1 to height

❖ d2 to depth

Then edit the Pro/program and set up the input section to read as follows:

```
input
width number
    "enter the width of the box"
    height number
    "enter the height of the box"
    depth number
    "enter the depth of the box"
end input
```

This will correctly prompt the user for the required values and assign them properly to the geometry when changes are made.

Incorrect Method

Setting up the input previously shown and editing the Pro/program *before* renaming the variables is incorrect. If you try to rename the variables once the

Pro/program is saved, the error message "Symbol already exists" appears. You can solve this problem one of two ways: (1) choose new names to assign to d0, d1, and d2 (e.g., width1, height1 and depth1) and then edit the Pro/program input section to use the same names, or (2) force the parameters to take on the values input by the user by adding the following in the relations section:

```
relations
    d0 = width
    d1 = height
    d2 = depth
    end relations
```

Turning Parts "Off" in an Assembly

Using Pro/PROGRAM in an assembly makes it possible to automatically ask users whether they would like components in the assembly. This can be easily accomplished by entering a question in the Pro/program input section. An example follows:

```
input
    Include_block  yes_no
    "do you want the block part in the assembly yes or no"
    end input
```

To tell Pro/ENGINEER how to perform the command, the Pro/program also has to include an If statement surrounding the Add Part statement:

```
if Include_block == YES

Add part block

. . . . . . . . . .

end add

end if
```

↬ **NOTE:** *If, when queried, the user inputs "no," Pro/ENGINEER will remove the part from the assembly. When the assembly is regenerated, the user will see "design contains warnings" in the message window. If you are editing the Pro/program you will see "warning: cannot find this model." Neither of the messages will affect program performance; internally, Pro/ENGINEER is trying to pass values to components not currently in memory.*

Common Errors

A list of other common, "simple" errors new users often make appears below:

❖ Deleting a parent feature that has children

❖ Reordering a child before its parent feature

❖ Inputting an incorrect spelling for a variable or input statement

❖ Missing an END IF statement following an IF statement

❖ Renaming a variable in a part which is "passed" via an execute statement from an assembly, without having the assembly in memory

❖ Forgetting to properly format comment lines using a slash and an asterisk (/*) at the beginning of each line

When a file contains an error, you can either abort or edit it. Abort cancels changes made to the program. Edit reinstitutes the editor window so that corrections can be made in the Pro/program itself.

↪ **NOTE:** *When using the JOT editor, choosing Edit can create an "infinite loop." The preceding section contains tips for avoiding and/or correcting this problem.*

Advanced Techniques

The parts and assemblies included with this book offer many examples of how to use advanced techniques. This section reviews the following basic functions:

❖ Execute statements

❖ Massprop statements

❖ If...Else clauses used for conditional feature/part suppression and/or logical statements

❖ Interact statements

Execute Statements

Execute statements input values from the top level assembly down *one level* at a time during assembly regeneration. Programs for each individual assembly part can be run independently, but Execute statements automatically force similar dimensions to be identical by "passing" input values from the assembly "down" into each part/subassembly.

➡ **NOTE:** *To "pass" a value down more than one level, place an Execute statement in the top level assembly to pass the value down to the subassemblies, and place Execute statements in the subassemblies to pass the value down to each part.*

A rule of thumb for adding Execute statements is to place them just before the Add Part or Add Sub-Assembly in the Pro/program. An example appears below:

```
input

   passed_value number

   "what is the value that you wish to pass to the parts?"

   end input


   relations

   .........

   end relations
```

```
execute part test1

passed_value_in_part = passed_value

end execute


add part test1

. . . . .

end add
```

The above code will take the input value (*passed_value*) from the top level assembly and assign it to the input parameter (*passed_value_in_part*) in the *test1* part. The same value can also be simultaneously passed to multiple parts or subassemblies. Of course, Pro/ENGINEER automatically regenerates all parts and subassemblies if the value changes.

➙ **NOTE:** *Although Execute statements only "pass" values down to input parameters in lower level parts and subassemblies, the initial value in the top level assembly does not itself have to be an input parameter. It could, for example, be calculated in a relation. However, Pro/ENGINEER will return an error if the input statement for the part or subassembly at which the Execute is directed does not contain the parameter. Chapter 7 presents additional examples of Execute statements and their functions.*

Massprop Statements

The final section in a Pro/program consists of Massprop statements. Entries made here force Pro/ENGINEER to automatically update the mass properties of parts or assemblies every time the Pro/program executes. The Massprop format appears below:

```
Massprop

Part          Name

Assembly      Name

End Massprop
```

For a thorough explanation of Massprop statements, see Chapter 8.

If...Else Clauses

If...Else logic can perform many functions within a Pro/program, including a few that Pro/PROGRAM itself does not allow. For example, if care is taken to account for all possible outcomes, If...Else logic can help capture design intent by mimicking "loops." Consider the following example:

```
input

   include_hole number

  "which hole feature do you want? 1=10mm, 2=20mm, 3=35mm, else 1"

   end input
```

```
relations

if include_hole == 3

hole_diameter = 35

else

if include_hole == 2

hole_diameter = 20

else

hole_diameter =10

endif

endif

end relations
```

The logic in the example will always return an answer; even if the user inadvertently enters a 4, the Pro/program should not crash or hang up.

If...Else logic can also be used to "release" a feature from the control of a relation. In effect, the geometry is manipulated by "overriding" the logic and directly modifying parameters. The following format accomplishes this task for the above example:

```
input

    include_hole number

    "which hole feature? 1=10mm, 2=20mm, 3=35mm, else override
relations"

    end input


    relations
```

```
if include_hole == 3

hole_diameter = 35

else

if include_hole == 2

hole_diameter = 20

else

if include_hole == 1

hole_diameter =10

else

endif

endif

endif

end relations
```

Entering anything other than 3, 2, or 1 causes the final Else command to "free" the relationship on *hole_diameter* and makes it possible for the user to manually modify the *hole_diameter* parameter.

Interact Statements

When inserted in the body of a Pro/program, an Interact statement temporarily suspends execution so that the user can insert more features in the part. Interact behaves like Insert mode in Pro/ENGINEER. A simple Interact statement appears below:

```
Add Cut.......
    If hole == "round"
```

```
        Add hole......
    Else
        Interact
    Endif
    Add Cut........
```

If the user does not enter "round," the Pro/program pauses and asks for an alternate set of features. The user enters one or more features, and Pro/PROGRAM replaces the Interact statement with the added features.

Summary

This chapter focused on advanced Pro/PROGRAM functionality, and introduced common errors encountered during program creation and their solution. With the exception of the Interact statement, all functionality described in this chapter is incorporated in *fan.asm*. Chapters 7 and 8 are dedicated to applying this functionality to real parts and assemblies.

Pulling It All Together Without Family Tables and Interchange Groups

This chapter covers stages in top level assembly creation using Pro/PROGRAM to make manipulation possible from any level, not only the top. The four stages of top level assembly creation follow:

1. Create standalone parts for the fan blade, motor, fan housings, switches, shipping containers, shipping foam, and the rest of the individual parts in *fan.asm*. (A complete parts list can be found in the Introduction.)

2. Create subassemblies for parts that ideally would arrive at the final assembly process pre-assembled (e.g., the motor and fan blade). Create interchange assemblies for swapping components such as switches.

3. Pull the parts and subassemblies into one top level assembly. Use this assembly to generate final round or box fans, create shipping foam block geometry, design the shipping box, calculate shipping cost, and estimate the final cost of the entire unit.

4. Create a drawing file of the top level assembly. Embed a dynamic table in the drawing file. Use the table to make changes to the assembly's controlling parameters and to report associated changes in shipping and manufacturing costs.

Stage 1: Advanced Part Creation

To demonstrate advanced part creation, let's focus on two parts as examples, the finger grill used on the box fan (*box_fan_grill.prt*) and the rear housing from the round fan (*rear_plastic_housing.prt*).

Box Fan Finger Grill

Box fan finger grill part (box_fan_grill.prt).

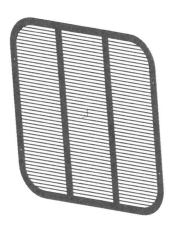

The box fan finger grill design requires 9 to 10 mm between the grill "rungs" and a modifiable flange around the outer perimeter. It must also size itself according to fan housing dimensions.

Corner detail.

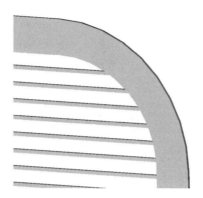

Simple Solution

A simple solution to the above design requirements would be to (1) create a solid part, (2) create a cut to form the "rung," and (3) pattern the cut to create the grill. These steps would work for a square-cornered part, but the finger grill has rounded corners. The problem with rounded corners is that one or more of the cuts will sometimes "cross" a tangent point between a rounded corner and straight edge, and sometimes (depending upon the size of the grill, number of cuts in the pattern, and so on) will not.

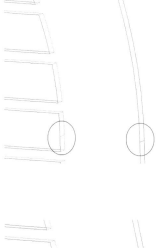

Detail of tangent point with cut spanning it.

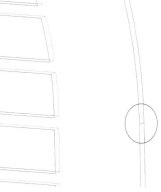

Detail of tangent point with cut not spanning it.

Another strike against this simple solution is that Pro/ENGINEER sometimes has difficulty when the number of edges intersected during the creation of a feature changes.

Alternate Solution

A better solution, albeit slightly more complex, follows.

1. Load *box_fan_grill.prt* into Pro/ENGINEER. Use Info | Regen Info to examine the Pro/program.

2. Create a protrusion larger than any conceivable fan housing size.

Large protrusion.

3. Create a cut and dimension it from the bottom edge of the housing. The dimension size should equal the desired flange width.

First cut with flange width.

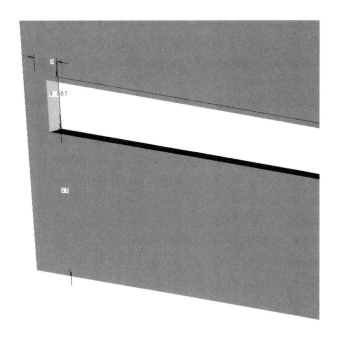

4. Develop relations for a pattern with the correct number of cuts and the correct flange width on top. Be sure to include the following: (1) The flange width on the top and bottom of the grill; (2) Between 9 to 10 mm of finger space between grill "rungs"; (3) Mounting holes located a quarter of the way down from the part top and a quarter of the way up from its bottom; and (4) a 50 mm radial clearance around the fan.

5. Create a small Pro/program to control inputs for the fan blade diameter, the grill depth, the grill flange width, the rib ("rung") height, the number of vertical stiffeners in the grill, and the corner radii value.

```
VERSION 16.0

REVNUM 11109

LISTING FOR GENERIC PART BOX_FAN_GRILL


INPUT

  FAN_DIA NUMBER

  "WHAT IS THE FAN BLADE DIAMETER?"

  GRILL_DEPTH NUMBER

  "ENTER THE DEPTH OF THE GRILL"

  GRILL_FLANGE_WIDTH NUMBER

  "ENTER THE WIDTH OF THE PERIMETER FLANGE FOR THE GRILL (RECO 25)"

  RIB_HEIGHT NUMBER

  "ENTER THE THICKNESS OF THE RIBS (RECO 2.5)"

  VERTICAL_STIFFENERS NUMBER

  "ENTER THE NUMBER OF VERTICAL STIFFENERS WANTED (RECO 2)"

  CORNER_RADII NUMBER

  "WHAT ARE THE CORNER RADII?"

END INPUT
```

Use the above inputs to set up relations for controlling the part geometry: (1) Set the box width and height equal to the fan diameter + 100 mm; (2) Create a variable to represent the height of the grill minus the top and bottom flange width. Call this variable "usable space"; (3) Calculate finger space *(usable space)* with the formula `(rib height*[pattern# - 1])/pattern#`. Pattern# equals the number of copies of the feature that exist in the pattern; and (4) Calculate the overall distance that each pattern travels with the formula, `rib height + finger space`.

The Pro/program calculation section for the patterns appears below.

```
RELATIONS
/*THIS SECTION SOLVES FOR THE NUMBER OF CUTS IN THE PATTERN TO
/*MAKE THE "GRILL" FEATURE. IT ASSUMES A NUMBER OF PATTERNED
/*CUTS, CALCULATES THE SIZE OF THE FINGER OPENING AND IF IT IS
/*LESS THAN 9MM IT REDUCES THE NUMBER OF CUTS BY 1. THIS
/*CONTINUES UNTIL IT REDUCES THE NUMBER UNTIL THE OPENING IS
/*BETWEEN 9 AND 10MM IN SIZE.
/*NOTE: IT WILL NEVER GO ABOVE 10MM SO LONG AS THE NUMBER OF RIBS
/*DOES NOT DROP BELOW 9.
D98 = CORNER_RADII
D96 = FAN_DIA + 100
D97 = D96
USABLE_SPACE = D96 - (2*GRILL_FLANGE_WIDTH)
D25 = GRILL_FLANGE_WIDTH
D118 = GRILL_FLANGE_WIDTH


P0 = 10
D22 = ((USABLE_SPACE - (RIB_HEIGHT*(P0-1)))/P0
IF D22 > 9
P0 = 80
D22 = ((USABLE_SPACE - (RIB_HEIGHT*(P0-1)))/P0
```

```
D26 = RIB_HEIGHT + D22

IF D22 < 9

P0 = 79

D22 = ((USABLE_SPACE - (RIB_HEIGHT*(P0-1)))/P0)

D26 = RIB_HEIGHT + D22

IF D22 < 9

P0 = 78

D22 = ((USABLE_SPACE - (RIB_HEIGHT*(P0-1)))/P0)

D26 = RIB_HEIGHT + D22

IF D22 < 9
```

Use the formulas in the last step to estimate rib height at 2 to 4 mm and pattern# at the maximum and minimum number of potential cuts. The results will be approximately 900 mm height for the largest fan housing and approximately 150 mm for the smallest. The results should indicate that 10 to 80 patterns will cover the 150 to 900 mm height range. Create the logic to have Pro/ENGINEER solve for pattern# based on fan size with the following steps:

1. Pro/ENGINEER assumes 80 patterns, uses the required rib height (user input), and the fan blade diameter (user input), and solves the above equations. If the result is a finger space of less than 9 mm, then the program assumes 79 patterns, solves the equations again, and continues to reduce the pattern count by one. The Pro/program stops stepping down the IF logic after the spacing passes 9 mm. The pattern spacing dimen-

sion is also solved for, the program stops when the finger space requirement is satisfied, and Pro/ENGINEER creates the cuts that determine proper flange height on the bottom of the grill.

Large "generic" grill with patterned cuts.

2. The grill part is completed with the use of the large "generic" grill. Create the cutout for the proper size grill, the perimeter flange, the mounting holes, and the vertical stiffeners by taking the large "generic" grill and making a cut representing the size and shape of the fan housing. Control the housing dimensions by inputting the desired fan blade diameter and the corner radii. This initial cut results in the intermediate grill.

Grill with initial cut.

3. To complete the grill flange, offset the curve inward to the flange thickness and extrude this section to the depth of the grill. Divide the grill's total width by the number of stiffeners needed, and create a protrusion to build the stiffener geometry.

*Grill with flange
and stiffeners.*

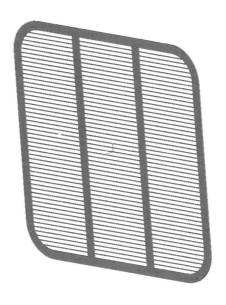

4. Complete the grill part by locating the holes at the center of the flange width, a quarter of the way down from the top, and a quarter of the way up from the bottom.

Completed grill.

Round Fan Rear Housing

Round fan rear housing part (rear_plastic_housing.prt).

Although the issues involved in configuring the round fan rear housing are similar to those for the box fan grill part, the solutions differ. As with the box fan grill, a key objective is to space fins ("rungs") so that fingers cannot touch the fan. The rear fan housing has two sets of fins, radial fins around the outside and concentric fins toward the center of the rear face.

Radial Fins

Close-up of radial fins.

The design of the radial fins requires 10 mm or less between fins (to prevent fingers from being able to pass through) and a flat area on top large enough for mounting the switches. The number of fins, the required distance between them, and the flat switch area are based on the outer diameter of the fan and the thickness of the plastic.

To automatically calculate the radial fins, use a rotational pattern and a number of relations. Start by creating the first member in the pattern, a thin revolved protrusion at an initial angle from vertical (feature number 7). The initial angle from vertical must be large enough to accommodate the switch area and calculated so that the area is centered relative to the number of radial fins.

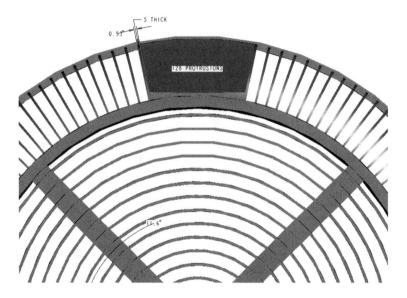

Thin revolved protrusion (feature number 7).

Assume an initial value for the initial angle (D476). Using this assumed value, and a calculation determining the proper rotation angle between fins based on the 10 mm fin distance requirement, calculate the number of fins required to go around the fan (variable CALC_P2). Round this value to the next largest integer. Using this value for the number of fins required, recalculate the initial angle from vertical and center the pattern from the vertical. Note that the RADIAL_FIN_PATTERN_ANGLE and the RADIAL_FIN_THICKNESS_ANGLE are based on the outer diameter of the housing (D479). Appearing below are the relations of *rear_plastic_housing.prt* showing calculations for radial fins.

```
/*Determine the outer case dia = fan dia + 45 mm clearance
OUTER_CASE_DIA=FAN_DIA+45


/*Dimension for the outer dia of the housing. Set to be a bit
/*smaller than the outer case dia for die draw, but still
/*allowing for good air flow.
D479=OUTER_CASE_DIA-5


/*Set up variables to use in calculations (C and D). This is done
/*so that the line length does not exceed 80 characters, and
/*entry of the relation is easier.
C=FIN_DISTANCE
D=PLASTIC_THICKNESS


/*Calculate the angle required to get proper distance between fins.
RADIAL_FIN_PATTERN_ANGLE=4*ACOS(((.5*SQRT(4*D479^2-(C+D)^2)-
    D479)/D479)+1)


/*Calculate the angle required to get the proper thickness of the
/*radial fins.
RADIAL_FIN_THICKNESS_ANGLE=4*ACOS(((.5*SQRT(4*D479^2-D^2)-
    D479)/D479)+1)
```

```
/*Initial angle from vertical for first radial fin. Set to this
/*value so that all the switches to be included will fit.
D476=12
```

```
/*Initial calculation of number of radial fins required in pattern.
CALC_P2=((360-2*D476)/RADIAL_FIN_PATTERN_ANGLE)+1
```

```
/*Force the value of P2 to the smallest integer not less than
/*CALC_P2.
P2=CEIL(CALC_P2)
```

```
/*Using this P2 value, recalculate the initial angle from vertical
/*for the first radial fin. This process will not allow it to
/*deviate very much from the initial value set above.
D476=(360-RADIAL_FIN_PATTERN_ANGLE*(P2-1))/2
```

```
/*Recalculate CALC_P2 using new values for the initial angle from
/*vertical for the first radial fin.
CALC_P2=((360-2*D476)/RADIAL_FIN_PATTERN_ANGLE)+1
```

➥ **NOTE:** *Using the above method, the flat area for mounting the switches always grows a little smaller from the initial value chosen. To make it always grow larger, choose the FLOOR command rather than the CEIL command to round down* *the number of required fins and generate a wider flat spot. However, neither method causes the value to fluctuate significantly from the initial value chosen.*

Pro/PROGRAM can set up many of the variables in this calculation for user input. However, "hardcoding" the variables makes change more difficult and design intent more transparent. FIN_DISTANCE, PLASTIC_THICKNESS, and the initial angle chosen for the flat for the switches (D476) indicate these advantages.

Concentric Fins

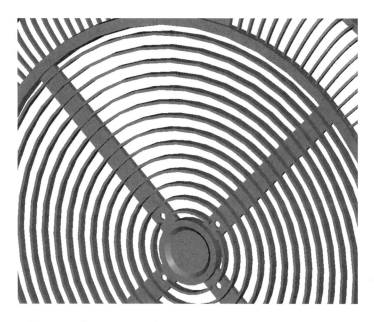

Close-up of concentric fins.

The concentric fins toward the center of the fan must be spaced at the 10 mm finger requirement, according to the plastic thickness, and relative to the inner diameter of the radial fins and the outer diameter of the center hub.

Start by creating the first fin in the concentric pattern. Create the pattern itself and use the radial dimension to control the distance between fins. Calculate the distance between fins (D716) using PLASTIC_THICKNESS and FIN_DISTANCE. Take the distance between the inner diameter of the radial fins and the outer diameter of the center hub and perform the calculation, CALC_P3, to compute the number of fins that will fit between them. Round this value down to an integer.

At this point and using the calculations shown below, center the pattern between the inner and outer diameters while maintaining the required 10 mm between fins. If 10 mm is not maintained, perform the calculation to center the fins again.

```
/*Calculate the diameters of the outer hub of the concentric
/*fins.
/*D532=FAN_DIA-65
D480=D532
D488=HOUSING_DEPTH-3
D715=D532-20


/*Calculate the dimension for the pattern distance between
```

```
/*concentric fins.

D716=FIN_DISTANCE+PLASTIC_THICKNESS


/*Initial calculation of number of concentric fins required

/*between inner and outer hubs.

CALC_P3=((D715-CENTER_HUB_DIA)/2-D716)/D716


/* Force the value of P3 to the largest integer not greater than

/*CALC_P3.

P3=FLOOR(CALC_P3)


/*Using this P3 value, the distance required to center the

/*pattern between the inner and outer hubs is calculated.

D713=(((D715-CENTER_HUB_DIA)/2)-D716*(P3-1))/2


/* Check to see if the dimension calculated would violate the min

/* fin distance.

IF D713>D716

    /* If so, add one member to the pattern, and recalculate D713.

    P3=P3+1

    D713=(((D715-CENTER_HUB_DIA)/2)-D716*(P3-1))/2

ENDIF
```

✓ **TIP:** *Both the radial fins and the concentric fins use an iterative calculation approach within the relations. Iterative calculations generally make very effective contributions to pattern automation.*

Stage 2: Subassemblies

A brief overview of the motor blade subassembly shows how the Pro/programs of individual parts and subassemblies can be integrated into a higher level assembly file. For information on the assemblies, see Chapters 3, 4, and 5.

blade_motor.asm Subassembly

Pro/PROGRAM can only pass values down one level at a time. Request all inputs at the top level and you can set up Pro/PROGRAM to "pass" these values down to parts and subassemblies, thereby guaranteeing component regeneration with the proper values.

The blade_motor.asm subassembly was created by combining *blades.prt* and *motor.prt*. Compare the input sections of the Pro/programs for the *blade_motor.asm* subassembly.

The Input section for *motor.prt* appears below.

```
INPUT
 MOTOR_DIA NUMBER
 "WHAT IS THE MOTOR DIA?"
 MOTOR_DEPTH NUMBER
 "WHAT IS THE MOTOR DEPTH?"
 SHAFT_DIA NUMBER
 "WHAT IS THE SHAFT DIA?"
END INPUT
```

The input section for the *blades.prt* follows.

```
INPUT

 BLADE_LENGTH NUMBER

 "ENTER THE LENGTH OF THE BLADES WANTED"

 BLADES NUMBER

 "ENTER THE NUMBER OF BLADES WANTED (3-10)"

 HUB_DEPTH NUMBER

 "ENTER THE DEPTH OF THE FAN BLADE HUB (30-40MM)"

 HUB_RADIUS NUMBER

 "ENTER THE HUB RADIUS (30-50MM)"

 SHAFT_OPENING_DIAMETER NUMBER

 "ENTER THE DIAMETER OF THE SHAFT OPENING"

END INPUT
```

The *blade_motor.asm* subassembly input section effectively combines the *motor.prt* and *blades.prt* input sections.

```
INPUT

 BLADES NUMBER

 "WHAT ARE THE NUMBER OF BLADES FOR THE FAN BLADE PART"

 HUB_DEPTH NUMBER

 "WHAT IS THE DEPTH OF THE BLADE - HUB"

 SHAFT_OPENING NUMBER

 "WHAT IS THE SHAFT SIZE OF THE MOTOR"

 HUB_RADIUS NUMBER

 "WHAT IS THE RADIUS OF THE HUB"
```

```
MOTOR_DEPTH NUMBER
```

"WHAT IS THE DEPTH OF THE MOTOR YOU WISH TO USE"

```
MOTOR_DIA NUMBER
```

"WHAT IS THE DIAMETER OF THE MOTOR YOU WISH TO USE"

```
FAN_DIA NUMBER
```

"WHAT IS THE DIAMETER OF THE FAN BLADES"

```
END INPUT
```

The relations section of the subassembly shows BLADE_LENGTH as determined by calculation using the fan diameter and hub radius.

```
RELATIONS
BLADE_LENGTH = (FAN_DIA - (2*HUB_RADIUS))/2
END RELATIONS
```

Pro/PROGRAM's Execute command "passes" the input values down to each of the subassembly's two parts. Execute statements from *blade_motor.asm* appear below. Some of the statements contain different variable names.

```
EXECUTE PART MOTOR
MOTOR_DEPTH = MOTOR_DEPTH
MOTOR_DIA = MOTOR_DIA
SHAFT_DIA = SHAFT_OPENING
END EXECUTE


ADD PART MOTOR
```

```
INTERNAL COMPONENT ID 9
END ADD


EXECUTE PART BLADES
BLADES = BLADES
HUB_DEPTH = HUB_DEPTH
BLADE_LENGTH = BLADE_LENGTH
SHAFT_OPENING_DIAMETER = SHAFT_OPENING
HUB_RADIUS = HUB_RADIUS
END EXECUTE


ADD PART BLADES
INTERNAL COMPONENT ID 10
PARENTS = 9(#5)
END ADD
```

In the preceding code block, the SHAFT_OPENING input value passes down to both the fan blade and motor parts.

✓ **TIP:** *Whenever possible, use the same names throughout the Pro/program at all levels. Although variable names in assemblies and parts do not have to be identical for the Execute command to work, using common names greatly reduces potential error.*

↝ **NOTE:** *Execute statements can only pass values to variables that exist within the input section of the* *subassemblies to which the values are passed. Saving the Pro/program for the motor part with MOTOR_DEPTH, MOTOR_DIA, and SHAFT_DIA missing from its Input section will generate a warning or an error when the assembly is regenerated.*

Stage 3: Top Level Assembly

The *blade_motor.asm* is a simple subassembly. The top level fan assembly (*fan.asm*), of course, is more complex. Created to integrate lower level components and subassemblies into a "complete" package, the top level assembly allows the user to make major changes without knowing the specifics of each component or subassembly. Changes at this level create a new design, complete with cost estimates. Imagine taking a request for a design change and generating a price quote in one phone call!

The top level assembly is governed by assumptions about practically every aspect of its design and creation: fan blades, switches, motors, and so on. For simplicity's sake the fan blade choices were limited to five sizes, from 500 to 750 mm in diameter. More than five unique designs are possible, however, because many parameters control the fan blades. Blade size alone does not necessarily equate to number of designs. The Input section of *fan.asm* showing fan blade choices appears next.

```
INPUT

 WHICH_BLADE NUMBER

 "WHICH BLADE 0=500MM, 1=600MM, 2=650MM, 3=700MM, 4=750MM"

 FAN_TYPE NUMBER

 "ENTER FAN CHOICE (1=BOX FAN, 2=ROUND FAN)"

 SWITCH_TYPE NUMBER

 "ENTER SWITCH CHOICE (1=ROCKER, 2=DIAL, 3=BUTTONS)"

 MOTOR_NUMBER NUMBER

 "ENTER THE MOTOR NUMBER (1 TO 5)"

 IF FAN_TYPE = 1

    GRILL_DEPTH NUMBER

    "ENTER THE DEPTH OF THE GRILL"

    GRILL_FLANGE_WIDTH NUMBER

    "ENTER THE WIDTH OF THE PERIMETER FLANGE FOR THE GRILL (RECO 25)"

    RIB_HEIGHT NUMBER

    "ENTER THE THICKNESS OF THE RIBS (RECO 2.5)"

    VERTICAL_STIFFENERS NUMBER

    "ENTER THE NUMBER OF VERTICAL STIFFENERS WANTED (RECO 2)"

    CORNER_RADII NUMBER

    "WHAT ARE THE CORNER RADII?"

 END IF

 PACKAGING YES_NO

 "WOULD YOU LIKE THE PACKAGING FOAM AND BOX SHOWN? (YES/NO)"

END INPUT
```

The following assumptions about the top level fan assembly were made in addition to the assumption for fan blade size:

❖ Two choices for fans: box and round

❖ Three choices for switches: rocker, dial, and button

❖ Five motors

❖ Grill specification for box fan

❖ Optional inclusion of shipping foam and box in the fan assembly and in costing

Box Fan vs. Round Fan

One of the first choices is between two types of fans, box or round, within the assembly.

Box fan.

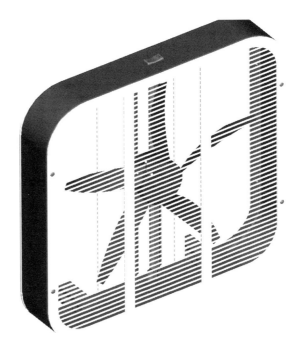

Round fan.

The two fan styles are radically different and are comprised of different components. The same top level assembly controls both, in part by using If loops around the Pro/program section that controls the components added to the assembly. As shown below, if you ask for a box fan (type 1), all parts required for the box fan are added to the assembly. If you ask for a round fan (type 2), the parts required to build the round fan are added.

```
IF FAN_TYPE==1

    EXECUTE PART BOX_FAN

    SWITCH_TYPE = SWITCH_TYPE

    FAN_DIA = FAN_DIA

    CORNER_RADII = CORNER_RADII

    BOX_FAN_DEPTH = BOX_FAN_DEPTH

    END EXECUTE


    ADD PART BOX_FAN

    INTERNAL COMPONENT ID 23

    END ADD

(many additional parts would be added here for box_fan)


ENDIF


IF FAN_TYPE==2

    EXECUTE PART BASE

    FAN_DIA = FAN_DIA
```

```
END EXECUTE


ADD PART BASE
INTERNAL COMPONENT ID 23
END ADD


ADD PART REAR_PLASTIC_HOUSING
INTERNAL COMPONENT ID 25
PARENTS = 23(#5)
END ADD
```

```
(many additional parts would be added here for round_fan)
ENDIF
```

Pro/PROGRAM "turns on" only the components needed for the assembly chosen. This function is useful for later operations as well, such as representing both assembly types with a single drawing.

✓ **TIP:** *Including the Execute statements required for each part or subassembly within the If loops lets Pro/ENGINEER pass only the values required by the upper level assembly. This saves valuable regeneration time, particularly when working with a large assembly.*

Multiple Switches within the Assembly

Using If loops as described in the preceding sections also makes it possible to properly add the components for the three different switch types to the top level assembly. Each switch assembly requires the addition of a different number of screws.

```
IF SWITCH_TYPE==1

    EXECUTE ASSEMBLY SW_ROCKER

    MATERIAL_TYPE = FAN_TYPE

    END EXECUTE


    ADD SUBASSEMBLY SW_ROCKER

    INTERNAL COMPONENT ID 15

    PARENTS = 12(*)

    END ADD


    ADD PART SWITCH_SCREW

    INTERNAL COMPONENT ID 260

    PARENTS = 15(*)

    END ADD
END IF


IF SWITCH_TYPE==2

    EXECUTE ASSEMBLY SW_DIAL

    MATERIAL_TYPE = FAN_TYPE

    END EXECUTE
```

```
(more parts and features)


END IF


IF SWITCH_TYPE==3

    EXECUTE ASSEMBLY SW_BUTTONS

    MATERIAL_TYPE = FAN_TYPE

    END EXECUTE


(more parts and features)


END IF
```

The bosses and housing cutouts also differ for each switch. To account for these differences in the top level assembly, the value of the switch is passed to the component to which it is mounted, and the Pro/program within the component adds or subtracts features as required. Appearing below is the *box_fan.prt* Pro/program showing If loops around features for switches.

```
IF SWITCH_TYPE==1


    ADD FEATURE (initial number 12)

    INTERNAL FEATURE ID 1811

    FEATURE WAS CREATED IN ASSEMBLY FAN
```

```
       PARENTS = 7(#3)


PROTRUSION: Extrude
(feature definition removed for clarity)


       ADD FEATURE (initial number 13)
       INTERNAL FEATURE ID 1882
       FEATURE WAS CREATED IN ASSEMBLY FAN
       PARENTS = 7(#3) 26(#6)
       END ADD
END IF


IF SWITCH_TYPE==2


       ADD FEATURE
       INTERNAL FEATURE ID 1959
       FEATURE WAS CREATED IN ASSEMBLY FAN
       PARENTS = 7(#3)


PROTRUSION: Extrude
(feature definition removed for clarity)


       ADD FEATURE
       INTERNAL FEATURE ID 2062
       FEATURE WAS CREATED IN ASSEMBLY FAN
       PARENTS = 7(#3) 26(#6)
       END ADD
```

```
END IF


IF SWITCH_TYPE==3


   ADD FEATURE

   (feature definition removed for clarity)

   END ADD
END IF
```

The creation of If loops requires more fore-thought than the average feature. Similar to family table instances (see Chapter 3), features cannot over-lap or exist simultaneously. For example, if all three switch cutouts in *box_fan.prt* exist at once, each attempts to cut away from the same piece of material, and this is problematic. To ensure that feature varia-tions do not coexist, create one feature while either manually or automatically suppressing the others.

To automatically suppress features according to component choice, use an input statement to assign a variable within the part and an Execute statement to pass it down from the top level assembly. For exam-ple, assign a variable SWITCH_TYPE as follows.

```
INPUT

 FAN_DIA NUMBER

 "ENTER FAN BLADE DIAMETER"

 SWITCH_TYPE NUMBER

 "ENTER SWITCH CHOICE (1=ROCKER, 2=DIAL, 3=BUTTONS)"
```

```
CORNER_RADII NUMBER

"ENTER THE CORNER_RADII OF THE HOUSING"

BOX_FAN_DEPTH NUMBER

"ENTER THE DEPTH OF THE BOX FAN HOUSING"

END INPUT
```

Once the variable is set and assigned to *box_fan.prt*, an If loop added around the features required to mount the switch allows creation only when SWITCH_TYPE is the chosen variable. Repeat this process for each switch type, and no top level duplications should occur during assembly regeneration. The same process applied to the round fan's *rear_plastic_housing.prt* achieves similar results.

☛ ***WARNING:*** *In assembly modeling, references to components and features must be present whenever the component or feature to which they refer is present. For example, the cutout for the rocker switch should not reference the geometry of the dial switch unless the dial switch geometry will be present when the rocker switch is chosen. If references are not present, the referenced items will not be present in the assembly and feature regeneration will be impossible.*

⤜ ***NOTE:*** *Using Pro/PROGRAM to regenerate only required features is particularly useful for accomplishing parts regeneration without family table instances.*

Five Fan Diameters and Five Different Motors

The number of fan diameters available to the assembly is five. Within the relations section of the assembly file, Execute statements pass the dimension variables assigned to build the *blades.prt* to the fan component. Appearing below is the Relations section of *fan.asm* showing the assignment of variables for *blades.prt*.

```
IF WHICH_BLADE == 0
   FAN_DIA = 500
   FAN_DEPTH = 30
   BLADES = 5
ENDIF
IF WHICH_BLADE == 1
   FAN_DIA = 600
   FAN_DEPTH = 35
   BLADES = 5
ENDIF
IF WHICH_BLADE == 2
   FAN_DIA = 650
   FAN_DEPTH = 35
   BLADES = 7
ENDIF
IF WHICH_BLADE == 3
   FAN_DIA = 700
   FAN_DEPTH = 40
```

```
    BLADES = 7

ENDIF

IF WHICH_BLADE == 4

    FAN_DIA = 750

    FAN_DEPTH = 40

    BLADES = 7

ENDIF
```

➴ **NOTE:** *For clarity and ease in incorporating design intent, the relation section also sets two other* blades.prt *variables, fan depth and number of blades. If desired, this type of variable assignment can also take place via user input statements.*

The five different fan motors are also assembled using relations to assign required variables for five unique *motor.prt* configurations. MOTOR_DIA and SHAFT_DIA are variables required by two parts, *motor.prt* and *blades.prt.* Execute statements ensure that these and other variables used by more than one part are appropriately passed to the parts that require them.

Pro/PROGRAM Inputs for Box Fan Grills

Users who choose a box fan are asked for specifications regarding its side grills. An If loop within the input section of the *fan.asm* containing all required parameters to recreate the *box_fan_grill.prt* responds to user specifications. Users respond only to requirement requests created by choices made earlier in the input list. For example, the *box_fan_grill.prt* inputs are not requested when the round fan is chosen, even if they are selected for modification during the regeneration. The If loop of the Input section for *fan.asm* follows.

```
IF FAN_TYPE ==1

    GRILL_DEPTH NUMBER

    "ENTER THE DEPTH OF THE GRILL"

    GRILL_FLANGE_WIDTH NUMBER

    "ENTER THE WIDTH OF THE PERIMETER FLANGE FOR THE GRILL (RECO 25)"

    RIB_HEIGHT NUMBER

    "ENTER THE THICKNESS OF THE RIBS (RECO 2.5)"

    VERTICAL _STIFFENERS NUMBER

    "ENTER THE NUMBER OF VERTICAL STIFFENERS WANTED (RECO 2)"

    CORNER_RADII NUMBER

    "WHAT ARE THE CORNER RADII?"

ENDIF
```

Packing Foam and Box

Packing foam and boxes are created for use only within the top level assembly. These components— *box_fan_foam.prt, box_fan_box.prt, round_fan_ foam.prt,* and *round_fan_box. prt*—are built in assembly mode using the techniques described Chapter 5. They reference other components within the fan assembly and, when the fan assembly is changed in any way, automatically regenerate to new configurations.

The packing foam and box dimensions vary according to the choices of inputs for the assembly. A 750 mm diameter fan, for example, requires a larger box and packing foam than a 500 mm diameter fan. Overall assembly depth can also change packaging size requirements such as when different motor and fan combinations are selected.

The top level assembly Input section contains a yes/no selection called Packaging for indicating whether or not to include the packaging foam and box within the assembly. The selection was created using a YES_NO input variable and If loops around the relevant Add Part sections of the *fan.asm* Pro/program. Upon selecting "no," the foam and box are not included in the assembly. Select "yes," and the foam and box are included.

```
PACKAGING YES_NO

"WOULD YOU LIKE THE PACKAGING FOAM AND BOX SHOWN? (YES/NO)"
```

The *fan.asm* program requires an If loop for each individual packing foam and box part for the box fan and round fan. To simplify the If loops for *fan.asm*, the program checks the yes/no status of Packaging, and then checks the type of fan requested. The *fan.asm* Pro/program showing the If loop around foam and box parts appears below.

```
IF PACKAGING==YES

    IF FAN_TYPE==1

        ADD PART BOX_FAN_FOAM

        INTERNAL COMPONENT ID 113

        END ADD


        ADD PART BOX_FAN_BOX

        INTERNAL COMPONENT ID 120

        END ADD

    END IF


    IF FAN_TYPE==2

        ADD PART ROUND_FAN_FOAM

        INTERNAL COMPONENT ID 117

        END ADD


        ADD PART ROUND_FAN_BOX

        INTERNAL COMPONENT ID 137

        END ADD

    END IF
```

The *foot.prt* component also plays a role in determining packaging size. Removing the feet for shipping reduces the required packaging space. If the feet are attached, the overall width of the fan increases and therefore requires a larger box and piece of foam.

Isometric view of fan.asm showing feet when fan is in use.

Isometric view of fan.asm showing feet in shipping position.

The problem presented by the *foot.prt* component is how to show the feet in proper placement when the packaging is not shown, and in a different placement when the packaging is shown. The solution is to assemble the feet in the assembly in both positions and use If loops to check whether the Packaging parameter is set to yes or no. If set to yes, the assembly includes the two feet in shipping position. If set to no, it includes the two feet in the use position. Either way, the bill of material is correct for the *fan.asm*, and the packaging for shipment is correct according to the position of the feet. The *fan.asm* Pro/program showing If loops around *foot.prt* of the box fan follows.

```
IF PACKAGING==NO

    ADD PART FOOT

    INTERNAL COMPONENT ID 111

    PARENTS = 14(#7) 12(#5)

    END ADD


    ADD PART FOOT

    INTERNAL COMPONENT ID 112

    PARENTS = 14(#7) 12(#5)

    END ADD

END IF


IF PACKAGING==YES

    ADD PART FOOT

    INTERNAL COMPONENT ID 126

    PARENTS = 12(#5)

    END ADD


    ADD PART FOOT

    INTERNAL COMPONENT ID 132

    PARENTS = 12(#5)

    END ADD

END IF
```

The logic of If loops also governs assembly for the screws that attach the switches to the fan case.

The If loop for each screw checks for the SWITCH_TYPE parameter.

The If loops for including or excluding feet and screws are added after all components have been assembled. If loops can span several components within an assembly, as is the case for the rocker switch attachment screws. The *fan.asm* Pro/program showing the If loop around the rocker switch (*sw_rocker.asm*) and the attachment screws (*switch_screw.prt*) appears below.

```
IF SWITCH_TYPE==1

   EXECUTE ASSEMBLY SW_ROCKER

   MATERIAL_TYPE = FAN_TYPE

   END EXECUTE


   ADD SUBASSEMBLY SW_ROCKER

   INTERNAL COMPONENT ID 15

   PARENTS = 12(#5)

   END ADD


   ADD PART SWITCH_SCREW

   INTERNAL COMPONENT ID 260

   PARENTS = 15(#8)

   END ADD


   ADD PART SWITCH_SCREW

   INTERNAL COMPONENT ID 261
```

```
    PARENTS = 15(#8)

    END ADD

END IF
```

☛ **WARNING:** *Assembly components must in no way reference items excluded from the assembly. This includes reference by geometry and features used to build or assemble components. If reference is made to excluded components, attempting to update the assembly components will cause a regeneration error or will freeze the component in place. Neither condition is acceptable. References to excluded components should be avoided at all costs.*

↝ **NOTE:** *References to excluded components are acceptable if and only if the component referencing the excluded component is itself excluded.*

Other Parameters Omitted from the Top Level Assembly Input List

Not all input statements required by subassemblies and parts must be contained within the input section of the top level assembly. Indeed, several parameters required by the lower level parts and subassemblies can be omitted. For example, the *box_fan.prt* requires a BOX_FAN_DEPTH input.

```
INPUT

 FAN_DIA NUMBER

 "ENTER FAN BLADE DIAMETER"

 SWITCH_TYPE NUMBER

 "ENTER SWITCH CHOICE (1=ROCKER, 2=DIAL, 3=BUTTONS)"

 CORNER_RADII NUMBER

 "ENTER THE CORNER RADII OF THE HOUSING"

 BOX_FAN_DEPTH NUMBER

 "ENTER THE DEPTH OF THE BOX FAN HOUSING"

END INPUT
```

However, the *fan.asm* Input section does not contain an input for BOX_FAN_DEPTH.

```
INPUT

 WHICH_BLADE NUMBER

 "WHICH BLADE 0=500MM, 1=600MM, 2=650MM, 3=700MM, 4=750MM"

 FAN_TYPE NUMBER

 "ENTER FAN CHOICE (1=BOX FAN, 2=ROUND FAN)"

 SWITCH_TYPE NUMBER

 "ENTER SWITCH CHOICE (1=ROCKER, 2=DIAL, 3=BUTTONS)"

 MOTOR_NUMBER NUMBER

 "ENTER THE MOTOR NUMBER (1 TO 5)"

 IF FAN_TYPE == 1

    GRILL_DEPTH NUMBER

    "ENTER THE DEPTH OF THE GRILL"

    GRILL_FLANGE_WIDTH NUMBER
```

```
"ENTER THE WIDTH OF THE PERIMETER FLANGE FOR THE GRILL (RECO 25)"

RIB_HEIGHT NUMBER

"ENTER THE THICKNESS OF THE RIBS (RECO 2.5)"

VERTICAL_STIFFENERS NUMBER

"ENTER THE NUMBER OF VERTICAL STIFFENERS WANTED (RECO 2)"

CORNER_RADII NUMBER

"WHAT ARE THE CORNER RADII?"

END IF

PACKAGING YES_NO

"WOULD YOU LIKE THE PACKAGING FOAM TO BE SHOWN? (YES/NO)"

END INPUT
```

Inputs not included in top level assembly input section can instead occur within top level assembly relations, and Execute statements can pass them to subassemblies and parts. Configuring inputs in this way enables someone who knows nothing of the lower level parts to manipulate the top level assembly and keep design intent intact.

The Relations section of *fan.asm* includes calculations to determine housing depth for both the box and the round fan. After they are calculated, the parameters BOX_FAN_DEPTH and ROUND_FAN_ HOUSING_ DEPTH are passed down to the housing component, which is then recreated according to the motor and blade choices entered at the assembly level. The Relations section of *fan.asm* showing calculation of housing depths follows.

```
BOX_FAN_DEPTH = FAN_DEPTH + MOTOR_DEPTH + 30

ROUND_FAN_DEPTH = FAN_DEPTH + MOTOR_DEPTH + 40

ROUND_FAN_HOUSING_DEPTH=ROUND_FAN_DEPTH/2
```

One switch parameter needed for assembly, MATERIAL_TYPE, passes down via Execute statements using another parameter, FAN_TYPE. The substitution of one parameter for the other is possible because an assumption for the material for each fan (metal for the box fan and plastic for the round) is the static value of its thickness. In this instance, using the Pro/program parameters simplifies the input section. The *fan.asm* Pro/program showing the Execute statement for the rocker switch follows.

```
EXECUTE ASSEMBLY SW_ROCKER

MATERIAL_TYPE = FAN_TYPE

END EXECUTE
```

➥ ***NOTE:*** *Avoid excessive simplification. For example, a parameter created equal to the thickness of the housing could pass to and change housings, switches, and other affected components at once. But, if the parameter changed, relations would have to change as well, and this change would not be automatic.*

Stage 4: Top Level Drawing

Once assembly creation is complete and the assembly works, a method is needed for retrieving information from the assembly in a suitable format. An assembly drawing typically accomplishes this task. This section illustrates an assembly drawing using the *cost.drw* example.

NAME	TYPE	MATERIAL	COST
FAN	ASSEMBLY	-	
BOX.FAN	PART	STEEL	3.420
BOX.FAN.GRILL	PART	PLASTIC	0.720
BOX.FAN.GRILL	PART	PLASTIC	0.720
SW_ROCKER	ASSEMBLY	-	
SW_ROCKER_HSG	PART	PLASTIC	0.150
SW_ROCKER_PIN	PART	STEEL	0.020
SW_ROCKER	PART	PLASTIC	0.050
SWITCH_SCREW	PART	STEEL	0.015
SWITCH_SCREW	PART	STEEL	0.015
BLADE.MOTOR	ASSEMBLY	-	
MOTOR	PART	-	0.700
BLADES	PART	PLASTIC	1.000
BOX.FAN_SCREWS	PART	STEEL	0.012
BOX.FAN_SCREWS	PART	STEEL	0.012
BOX_FAN_SCREWS	PART	STEEL	0.012
BOX.FAN_SCREWS	PART	STEEL	0.012
BOX.FAN_SCREWS	PART	STEEL	0.012
BOX.FAN_SCREWS	PART	STEEL	0.012
BOX.FAN_SCREWS	PART	STEEL	0.012
BOX.FAN_SCREWS	PART	STEEL	0.012
BOX.FAN_VERT.MEMBERS	PART	STEEL	0.147
BOX.FAN_VERT.MEMBERS	PART	STEEL	0.147
MOTOR.SCREW	PART	STEEL	0.010
MOTOR.SCREW	PART	STEEL	0.010
MOTOR.SCREW	PART	STEEL	0.010
MOTOR.SCREW	PART	STEEL	0.010
FOOT	PART	PLASTIC	0.250
FOOT	PART	PLASTIC	0.250
LABOR.COST	BULK ITEM	-	1.250
OVERHEAD	BULK ITEM	-	5.000
		TOTAL COST	13.991

SCALE 1:3

SCALE 1:8

X.X   +-0.1
X.XX   +-0.10
X.XXX   +-0.100
ANG.   +-1.0

REP : Master Rep
SCALE : 1:3   TYPE : ASSEM   NAME : FAN   SIZE : D   SHEET 1 OF 2

Example cost.drw drawing.

Assembly views such as isometric, front, side, and rear help the user to visualize results and allow verification of choices made during regeneration.

✓ **TIP:** *Simplify views using the No Hidden and Tan Dimmed commands under Views | Display Mode | View Display. This is especially effective for assemblies and isometric views.*

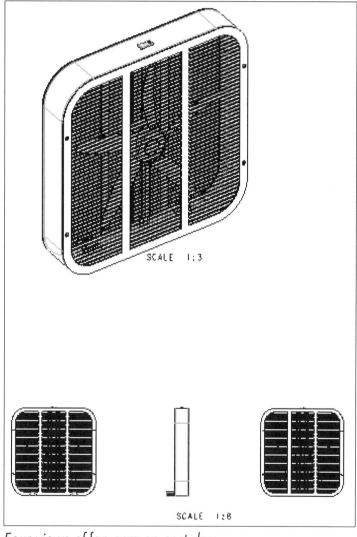

Four views of fan.asm *on* cost.drw.

Information required of the assembly includes bill of material, cost of manufacturing, shipment packaging requirements, and component material composition. Changes made to the assembly should cause this information to update dynamically. A Pro/ENGINEER functionality called Drawing Tables creates dynamic tables for presenting assembly information. The Pro/ENGINEER module, Pro/REPORT, retrieves the information and presents it in customized reports.

➥ **NOTE:** *Much of what this section covers is possible only with the Pro/REPORT module.*

The process for creating a table is simple. Use the Table | Create commands to input the desired number of columns and rows. The table for *cost.drw* consists of three rows and four columns for component name, type, material, and cost.

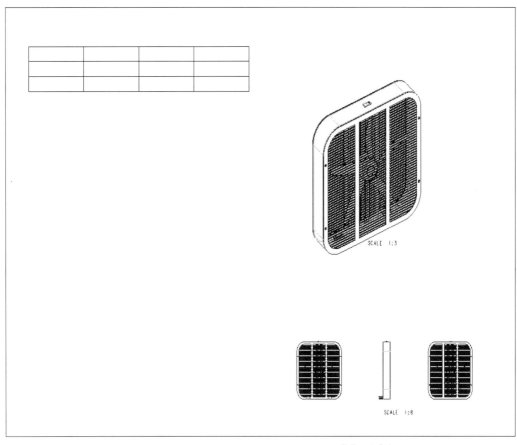

SCALE 1:3

SCALE 1:8

X.X   +-0.1
X.XX +-0.10
X.XXX +-0.100
ANG. +-1.0

REP : Master Rep
SCALE : 1:3   TYPE : ASSEM NAME : FAN SIZE : D SHEET 1 OF 2

Created table in cost.drw.

✓ **TIP:** *Columns and rows may be added and sub-tracted at any time using the Table | Mod Rows/ Cols commands. However, planning ahead in an attempt to initially create the exact number of rows and columns required is recommended.*

Adding or subtracting rows or columns is tricky and can have unpredictable effects on the table.

Add text to the table using the Table | Enter Text commands. Enter text with the keyboard or choose report symbols via Report Sym. Report symbols make drawing tables dynamic.

The initial information entered in a table is not specific to any model within the assembly nor required to change with any model. In essence, the pieces of information act as column headings. Use the keyboard to input this information. For *cost.drw*, enter the text NAME in the first column, TYPE in the second, MATERIAL in the third, and COST in the fourth.

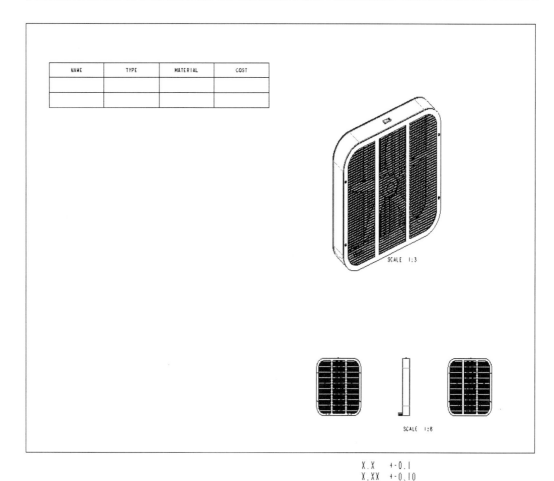

NAME	TYPE	MATERIAL	COST

SCALE 1:3

SCALE 1:8

X.X   +-0.1
X.XX   +-0.10
X.XXX +-0.100
ANG.   +-1.0

REP : Master Rep
SCALE : 1:3   TYPE : ASSEM NAME : FAN SIZE : D SHEET 1 OF 2

Drawing table showing initial text created.

A repeat region is used to create the table's dynamic portion. Repeat regions tell tables whether to expand or contract to meet information changes within their associated assemblies. Repeat regions can be simple, two-dimensional, or nested.

Tables with simple repeat regions expand in only one direction. Add components to an assembly whose table contains a simple repeat region, and the region expands the table. Remove components and the region contracts the table.

Two-dimensional (2D) repeat regions expand tables two ways and are useful for documenting information such as the data contained in family table instances. Family tables change according to component modifications as well as changes in dimension values, features, and parameters. A table with a 2D repeat region will expand and contract automatically whenever changes in the family table occur. A 2D repeat region can also show all columns and rows contained in the family table.

Nested repeat regions are entirely contained by other repeat regions. Nesting repeat regions make possible tasks such as searching the subassemblies of an assembly and the parts of a subassembly.

➥ **NOTE:** *Take care to select the proper repeat region type prior to table creation, and investigate potential benefits and problems prior to selection. The* Pro/ENGINEER *Drawing User's Guide shipped with the Pro/ENGINEER software discusses repeat region types in depth.*

The *cost.drw* table is simple, requiring only one direction of expansion to show all assembly components; therefore, a simple repeat region will suffice.

The second row, first column is the repeat region's upper left corner, and the second row, fourth column is its right lower corner. Report symbols entered in this region will expand vertically across columns and automatically include all assembly members.

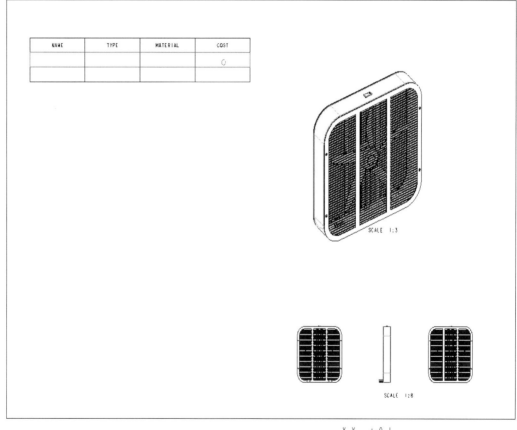

Drawing table showing repeat region selected.

Select the report symbols to use for expanding the assembly and place them in the repeat region using the Table | Enter Text | Report Sym commands. A cell within a repeat region must be selected to use Report Sym functionality. In *cost.drw,* component names, types, materials, and costs are all responsive to assembly expansion.

Component name and type are included in Pro/REPORT as *asm.mbr.name* and *asm.mbr.type.* Because they are system parameters, they can be directly selected from the Report Sym list. For information on additional system parameters, see the *Pro/ENGINEER Drawing User's Guide.*

Component material and cost are user-defined parameters and must be entered manually. Select *asm.mbr* from the Report Sym list, choose User Defined, and input the symbol text for the desired parameter. To input the *cost.drw* parameters, type MATERIAL and COST.

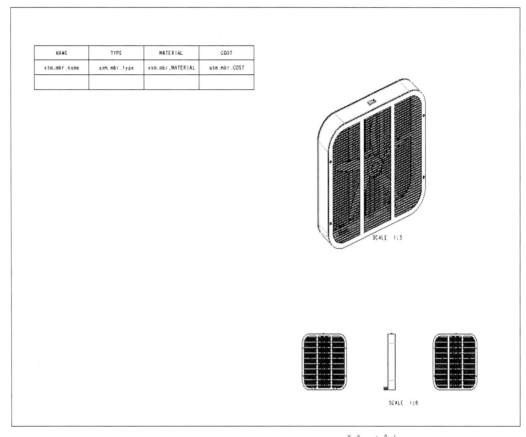

NAME	TYPE	MATERIAL	COST
asm.mbr.name	asm.mbr.type	asm.mbr.MATERIAL	asm.mbr.COST

SCALE 1:3

SCALE 1:8

X.X   +-0.1
X.XX   +-0.10
X.XXX   +-0.100
ANG.   +-1.0

REP : Master Rep
SCALE : 1:3   TYPE : ASSEM   NAME : FAN SIZE : D SHEET 1 OF 2

Drawing table with text in repeat region.

Use the Table | Repeat Regions | Update Tables commands to cause repeat region updating for current assembly status, and recalculation of the parameters for current values. Using Draft to regenerate also causes recalculation to occur.

NAME	TYPE	MATERIAL	COST
BOX_FAN	PART	STEEL	3.420
BOX_FAN_GRILL	PART	PLASTIC	0.720
BOX_FAN_GRILL	PART	PLASTIC	0.720
SW.ROCKER	ASSEMBLY	·	
SWITCH_SCREW	PART	STEEL	0.015
SWITCH_SCREW	PART	STEEL	0.015
BLADE_MOTOR	ASSEMBLY	·	
BOX_FAN_SCREWS	PART	STEEL	0.012
BOX_FAN_SCREWS	PART	STEEL	0.012
BOX_FAN_SCREWS	PART	STEEL	0.012
BOX_FAN_SCREWS	PART	STEEL	0.012
BOX_FAN_SCREWS	PART	STEEL	0.012
BOX_FAN_SCREWS	PART	STEEL	0.012
BOX_FAN_SCREWS	PART	STEEL	0.012
BOX_FAN_SCREWS	PART	STEEL	0.012
BOX_FAN_VERT_MEMBERS	PART	STEEL	0.147
BOX_FAN_VERT_MEMBERS	PART	STEEL	0.147
MOTOR_SCREW	PART	STEEL	0.010
MOTOR_SCREW	PART	STEEL	0.010
MOTOR_SCREW	PART	STEEL	0.010
MOTOR_SCREW	PART	STEEL	0.010
FOOT	PART	PLASTIC	0.250
FOOT	PART	PLASTIC	0.250
LABOR_COST	BULK ITEM	·	1.250
OVERHEAD	BULK ITEM	·	5.000

SCALE 1:3

SCALE 1:8

```
X.X    +-0.1
X.XX   +-0.10
X.XXX  +-0.100
ANG.   +-1.0
```

REP : Master Rep
SCALE : 1:3   TYPE : ASSEM NAME : FAN SIZE : D SHEET 1 OF 2

Drawing table with repeat region updated to current values.

User-defined parameters may or may not be defined in individual models at the time of table creation. For example, if the MATERIAL and COST parameters are defined within the component, their current values display automatically. If they are not defined, the columns that correspond to them remain blank.

Using the standard Add Param command is one way to define undefined component parameters. Parameter definition can also take place in a relation within the component. A third way is to use the MODIFY command, picking directly on the table and then entering the parameter value. If the value is a number, it can be entered directly. If the value is a string, it should be included within quotes.

Repeat Region Attributes

Repeat region attributes control how regions expand relative to the assembly structure. The attributes Recursive and Flat, for example, control parameter searches. Recursive searches all levels of the data for the parameters specified in the repeat region, and produces a report of the assembly structure similar to an assembly tree. The Flat attribute searches only the top level of the data, producing reports for subassembly names and assembly components. However, Flat reports do not include names of components within subassemblies.

The attributes Duplicates, No Duplicates, and No Dup/Level control handling of duplicate parameter occurrences. The Duplicate attribute causes duplicate occurrences of a parameter to list separately within the repeat region. No Duplicates causes duplicate parameters to list only once. If &*rpt.qty* is specified, the No Duplicates attribute will display values along with duplicate parameters. No Dup/Lev also lists duplicate parameters within each level once, but lists the parameter at each level where it occurs. As with No Duplicates, specifying &*rpt.qty* lists each level's duplicate quantity.

For a screw used within both an assembly and an associated subassembly, duplicate attributes function as follows:

❖ *Duplicates* reports each screw used as an individual line item.

❖ *No Duplicates* reports one line item and displays the quantity used for the assembly and subassembly in the quantity column.

❖ *No Dup/Lev* lists the screws used in the top level assembly as one line item, the screws used in the subassembly as another line item, and indicates the quantities used at each level.

To set repeat region attributes, use the Table | Repeat Region | Attributes commands.

Relations within Repeat Regions

To write and display parameter relations within a repeat region (e.g., to calculate the total cost of a component by multiplying the number of items it contains by individual cost), select Table | Repeat Region | Relations. Note that relations created within a repeat region also store within the region; they cannot be referenced outside the region.

Summation of Parameter Values within Repeat Regions

Pro/ENGINEER can automatically add all individual values for a parameter and display that sum as a total in the table. To accomplish this, add a summation parameter to the repeat region using the Table | Repeat Region | Summation commands, select the variable to be summed, and enter a parameter name to identify the sum value. Include the sum parameter in cells within the table or as a note using &*parametername*.

•• **NOTE:** *You must use Update Tables to update summation parameters.*

☞ **WARNING:** *Summing a parameter within a region attributed as No Duplicates or No Dup/Lev will cause errors. In the presence of these attributes, Pro/ENGINEER will not factor item quantity in its calculations.*

For example, in an assembly with four screws and a subassembly with two, each screw costs $0.05. With Duplicates attributed, each screw lists separately, and Pro/ENGINEER sums cost at $0.30. With No Duplicates attributed, the cost result is $0.05. With No Dup/Lev, the sum, $0.10, or $0.05 per level, is again incorrect.

Indentation Within Repeat Regions

Indentation within repeat regions is made possible by setting attributes to recursive. Select Table | Repeat Region | Indentation and enter the indentation amount according to the number of characters. Set indentation to cause subassemblies and components to indent as they would in an assembly tree. This is a useful format for parameters such as *asm.mbr.name.*

Merging and Remeshing Cells Within A Table

Pro/ENGINEER's cell merging and splitting functions appear under Table | Modify Table as Merge and Remesh. Both work well for creating readable, usable tables.

Saving and Restoring Tables

To apply one table to more than one product, use Pro/ENGINEER'S Table | Save/Retrieve commands for saving and restoring drawing tables. The command functionality is explained clearly during use.

Multiple Models within Drawings with Tables

Drawings with tables can contain multiple models. The tables take on the values of and attach to whichever model is current when the table is created.

Overview

Take the following steps to prepare a completed table for *fan.asm:*

1. Add indentation to the first column.

2. Apply the attributes Duplicates and Recursive.

3. Add a summation called SUM_COST and place it in the last row, fourth column.

4. Add all required parameter and relation information to all components in the assembly.

5. Merge the cells for the last row, second and third columns, and add the text, "TOTAL COST."

NAME	TYPE	MATERIAL	COST
FAN	ASSEMBLY	-	
BOX.FAN	PART	STEEL	3.420
BOX.FAN.GRILL	PART	PLASTIC	0.720
BOX.FAN.GRILL	PART	PLASTIC	0.720
SW.ROCKER	ASSEMBLY	-	
SW.ROCKER.HSG	PART	PLASTIC	0.150
SW.ROCKER.PIN	PART	STEEL	0.020
SW.ROCKER	PART	PLASTIC	0.050
SWITCH.SCREW	PART	STEEL	0.015
SWITCH.SCREW	PART	STEEL	0.015
BLADE.MOTOR	ASSEMBLY	-	
MOTOR	PART	-	0.700
BLADES	PART	PLASTIC	1.000
BOX.FAN.SCREWS	PART	STEEL	0.012
BOX.FAN.SCREWS	PART	STEEL	0.012
BOX.FAN.SCREWS	PART	STEEL	0.012
BOX.FAN.SCREWS	PART	STEEL	0.012
BOX.FAN.SCREWS	PART	STEEL	0.012
BOX.FAN.SCREWS	PART	STEEL	0.012
BOX.FAN.SCREWS	PART	STEEL	0.012
BOX.FAN.SCREWS	PART	STEEL	0.012
BOX.FAN.VERT.MEMBERS	PART	STEEL	0.147
BOX.FAN.VERT.MEMBERS	PART	STEEL	0.147
MOTOR.SCREW	PART	STEEL	0.010
MOTOR.SCREW	PART	STEEL	0.010
MOTOR.SCREW	PART	STEEL	0.010
MOTOR.SCREW	PART	STEEL	0.010
FOOT	PART	PLASTIC	0.250
FOOT	PART	PLASTIC	0.250
LABOR.COST	BULK ITEM	-	1.250
OVERHEAD	BULK ITEM	-	5.000
		TOTAL COST	13.991

SCALE 1:3

SCALE 1:8

X.X   +-0.1
X.XX   +-0.10
X.XXX +-0.100
ANG.   +-1.0

REP : Master Rep
SCALE : 1:3   TYPE : ASSEM   NAME : FAN SIZE : D SHEET 1 OF 2

Completed table for fan.asm.

The table for *fan.asm* thus created is dynamic and will update with the assembly.

Engineering Notebooks: Another Use of Drawing Tables

Another use of drawing tables is as an upfront interface to the user, sometimes called an "engineering notebook." Instead of instructing Pro/ENGINEER to prompt the user for input values during a regeneration, a table may be set up that contains all input parameters and respective values. These values can then be modified simply by picking on the table and inputting the new values.

INPUT	CURRENT VALUE
WHICH_BLADE	0
FAN_TYPE	2
SWITCH_TYPE	1
MOTOR_NUMBER	1
PACKAGING	FALSE
IF FAN_TYPE IS = 1, ENTER THE FOLLOWING	
GRILL_DEPTH	2.50
GRILL_FLANGE_WIDTH	25.00
RIB_HEIGHT	2.50
VERTICAL_STIFFENERS	2
CORNER_RADII	100.00

VIEW OF CURRENT
STATUS OF FAN.ASM

SCALE 1:3

X.X   +-0.1
X.XX   +-0.10
X.XXX +-0.100
ANG. +-1.0

SCALE : 1:3   TYPE : ASSEM NAME : FAN SIZE : D SHEET 2 OF 2

Sheet 2 of cost.drw.

The creation of a drawing table that fulfills this function is rather simple. One of the two methods discussed below is automated and uses Pro/REPORT. The other method is manual. The example used in this section is Sheet 2 of the *cost.drw* file.

To initially create the drawing table, use the same Table Create commands employed in the previous section. In the example, the intent is to list only the variable names and their values. Consequently, only a two-column table is required.

However, depending on whether you plan to use Pro/REPORT and repeat regions to automatically query the file for all input parameters, or fill the table manually with only the parameters you want, the number of rows you may wish to use differs. For the automatic method, only a couple of rows are necessary because Pro/REPORT will automatically fill in the repeat region. For the manual method, you need enough rows to accommodate all parameters you want to include in the table. As stated previously, you can change the number of columns and rows at any time using Table | Mod Rows/Cols commands, but initial planning can save you many hassles later on.

The automatic method requires that you create a repeat region, and populate it with text using the Report Sym command. The proper report symbols are *mdl.param.name* and *mdl.param.value*. Note that this method has certain advantages and disadvantages. If parameters are added and removed from the *fan.asm*, the table will automatically update.

However, this can also be a problem. The report symbols will add all parameters within the assembly to the table. Unless a proper filter can be found that will display only the parameters that the user would like displayed in the table, there will be many unnecessary parameters in the table.

✓ *TIP: Using a unique character or combination of characters within all input parameters can be an effective method of setting up a filter. If the character or combination is used only within inputs, you can easily set up a filter with the following form:* &mdl.param.name==*unique character combination*. *The former will filter the repeat region and return only the parameter names that contain the unique character combination. An example might be to include a double underscore (__) within all parameters that will be used as inputs. Then a filter can be set up with the* &mdl.param.name==*__* *form.*

The *fan.asm* example on the companion CD includes 20 parameters in the assembly. Only 10 of the parameters are used as inputs; the other 10 are calculated within relations or set as parameters using the Relations | Add Param commands. If the parameters were to be left within the table, it would be extremely confusing for the user to determine which parameters s/he would have to change to modify the assembly.

An easier method is to manually input the parameter names and values by entering the corresponding symbol in the table. This method provides the user with full control over the parameters to be included in the table. To accomplish this task, use the Enter Text | Keyboard commands in the TABLE menu.

The parameter names are entered directly as regular text. In the example, the WHICH_BLADE input shown at the top of the table is input directly as "WHICH_BLADE." There is no direct connection to the actual parameter name; if a parameter name changes, the text will not update.

The parameter values are entered by directly inserting the symbol for the corresponding parameter. In the example, the text entered for the WHICH_BLADE input would be &WHICH_BLADE. Using the symbol tells Pro/ENGINEER to return the value for the parameter. This method was used to create the table shown on Sheet 2 of the *cost.drw* drawing file. Limitations are evident in that the table does not dynamically change with updates to assembly parameters.

Table on Sheet 2 of cost.drw.

INPUT	CURRENT VALUE
WHICH_BLADE	0
FAN_TYPE	2
SWITCH_TYPE	1
MOTOR_NUMBER	1
PACKAGING	FALSE
IF FAN_TYPE IS = 1, ENTER THE FOLLOWING	
GRILL_DEPTH	2.50
GRILL_FLANGE_WIDTH	25.00
RIB_HEIGHT	2.50
VERTICAL_STIFFENERS	2
CORNER_RADII	100.00

Either method creates a table which can then be used by Pro/ENGINEER neophytes to modify the controlling parameters of the assembly. Modification of the parameters is as simple as choosing the MOD-IFY command, selecting the value you wish to change, and inputting the new value. Regeneration can then occur using the Current Values command because the parameters have already been changed. One limitation of this method is the loss of a line prompt that appears in the Message window when entering new values while regenerating the assembly. A partial workaround is to properly add text in the table that defines every parameter, and specifies the range of the selected value. The manually entered text would not update with changes to the Pro/program.

Exploring drawing table and repeat region functionality takes time and is best accomplished on a product by product basis. Neither function is particularly complex. Practice, trial, and error are the best means for determining the value of either to a given project.

Summary

This chapter focused on the four stages used in creating a top level assembly with Pro/PROGRAM. The stages include the creation of many standalone parts, some of which contained advanced part creation techniques; subassemblies; a top level assembly; and a dynamic drawing file that contains tables allowing changes to the values of the assembly's controlling parameters, as well as reporting associated changes in shipping and manufacturing costs. The top level assembly in this chapter was created while ignoring family tables and interchange groups. The latter topics and how they change the process are discussed in Chapter 8.

Pulling It All Together with Family Tables and Interchange Groups

In Chapter 7 all parts/subassemblies were fully integrated into a single top level assembly called *fan.asm,* and an integrated drawing was created that allowed even a novice to evaluate hundreds of different fan combinations. However, the top level assembly and Pro/program in Chapter 7 used neither family tables nor interchange groups.

In this chapter an alternate top level assembly called *alt_box_fan.asm* will be created. For simplicity's sake, this assembly consists of the box fan geometry only and does not include packing foam, packing box, or feet in packing position.

*Assembly
alt_box_fan.asm.*

As in Chapter 7, the *alt_box_fan.asm* assembly is a collection of many smaller standalone assemblies and parts. The term "standalone" means that every part or subassembly comes with its own unique Pro/

program. This allows easy manipulation of the files at any level within the organization. You are not forced to make changes from the top level assembly only.

The sections that follow show how to create a Pro/program that utilizes all of the box fan geometry and incorporates family tables and interchange groups. The focus is on the changes required, not on the kind of detailed walkthrough required for the top level assembly in Chapter 7.

Because family tables and interchange groups require different dimensioning schemes, an alternate assembly, *alt_blade_ motor.asm,* will be created along with four alternate parts—*alt_motor.prt, alt_blades.prt, alt_ box_fan_vert_ members.prt,* and *alt_box_fan_ housing.prt.*

Initial Subassembly

As mentioned previously, Pro/PROGRAM can only pass values down one level at a time in an assembly. The most powerful way to use Pro/PROGRAM is to use each successive subassembly level as a means to consolidate all required inputs to control the lower parts and subassemblies. Request all inputs at the top level, and you can set up Pro/PROGRAM Execute statements that will "pass" values down to all parts and subassemblies simultaneously. This procedure guarantees that all components will regenerate with the proper values.

To create the *alt_blade_motor.asm* subassembly, combine *alt_blades.prt* and *alt_motor.prt*. Pass two parameters from the subassembly down to the *alt_blades.prt*, BLADES, and FAN_DIA. The Execute statement from the *alt_blade_motor.asm* subassembly appears below.

```
EXECUTE PART ALT_BLADES
BLADES = BLADES
FAN_DIA = FAN_DIA
END EXECUTE


ADD PART ALT_BLADES
INTERNAL COMPONENT ID 17
PARENTS = 18(#5)
END ADD
```

✓ **TIP 1:** *To reduce the potential for error, use the same names whenever possible throughout the program and at all levels.*

✓ **TIP 2:** *Including the Execute statements required for each part or subassembly within the If loops allows Pro/ENGINEER to pass the values to the part or subassembly only when required by the upper level assembly. This saves valuable regeneration time, particularly when working within a large assembly.*

➡ **NOTE:** *Execute statements can only "pass" values to variables that exist within the Input section of the parts and subassemblies to which the values are being passed.*

Top Level Assembly

The top level *alt_box_fan.asm* assembly fully integrates all work done on the lower level components and subassemblies into a "complete" package. This package allows you or another user to make major changes to the assembly without having to know the specifics of each component or subassembly.

Several assumptions are made for *alt_box_fan.asm*. As in Chapter 7, fan blade choices are limited to five sizes ranging from 500 to 750 mm in diameter. The Input section of *alt_box_fan.asm* showing fan blade choices follows.

```
INPUT

 WHICH_BLADE NUMBER
 "WHICH BLADE 0=500MM, 1=600MM, 2=650MM, 3=700MM, 4=750MM"
```

Additional assumptions are listed below:

❖ Three switch choices: rocker, dial, and button

❖ Five possible motors

❖ User-defined grills

Assumptions are made for *alt_box_fan.asm* in order to make following the demonstration in this chapter easier. Pro/ENGINEER offers dozens of solutions for any given problem. These assumptions allow you to explore different alternatives within the same top level assembly. For simplicity of demonstration, several assembly parameters have been omitted from the Input list. These parameters are described later in the chapter.

Multiple Switches within the Assembly

The three switch alternatives are assembled into the top level box fan using an interchange group. (Chapter 4 demonstrates how to create the switches interchange group.) To tell Pro/ENGINEER how to select between the switches, assemble one member (part or subassembly) of the group into the top level and type in the number assigned in the Input section of the Pro/program.

```
INPUT

SWITCH_TYPE NUMBER
"ENTER SWITCH CHOICE (1=ROCKER, 2=DIAL, 3=BUTTONS)"
```

To indicate a part or subassembly preference using a relation, assign the interchange assembly name associated with the proper switch subassembly to a variable. In the example, the variable name is SWITCH_NAME. The assignment of the name string in the Relations section follows:

```
IF SWITCH_TYPE==1

   SWITCH_NAME = "SW_ROCKER.ASM"

   ELSE

   IF SWITCH_TYPE==2

      SWITCH_NAME = "SW_DIAL.ASM"

      ELSE

      IF SWITCH_TYPE==3

         SWITCH_NAME = "SW_BUTTONS.ASM"

      ENDIF

   ENDIF

ENDIF
```

➥ **NOTE:** *Yet another way to specify component preference is to type in the full string of the component name. This method is used below for motor selection with family tables.*

After assigning the name to a variable, edit the ADD statement in the top level assembly to enclose the variable name in parentheses. Pro/ENGINEER now recognizes the variable as an interchange group.

```
ADD COMPONENT (SWITCH_NAME)

INTERNAL COMPONENT ID 131

PARENTS = 22(#6)

END ADD
```

Making an interchange group functional requires several additional steps. For the switches group, the first step is to build the rest of the assembly using If loops to supply the features and components required by each switch such as screws, bosses, and housing cutouts. For example, a Pro/program in the housing part will add or subtract features according to the switch value it receives. The *alt_box_fan_housing.prt* Pro/program showing If loops around switch features appears below.

```
NOTE: Many lines removed for clarity....


IF SWITCH_TYPE==1


   ADD FEATURE (initial number 17)
   FEATURE WAS CREATED IN ASSEMBLY ALT_BOX_FAN
PROTRUSION: Extrude
SLOT: Extrude
   END ADD
END IF


IF SWITCH_TYPE==2


   ADD FEATURE (initial number 17)
   FEATURE WAS CREATED IN ASSEMBLY ALT_BOX_FAN
PROTRUSION: Extrude
SLOT: Extrude
   END ADD
END IF
```

```
IF SWITCH_TYPE==3


    ADD FEATURE
    FEATURE WAS CREATED IN ASSEMBLY ALT_BOX_FAN
PROTRUSION: Extrude

SLOT: Extrude

    END ADD
END IF
```

Creation of the If loop requires more forethought than the average feature. As with family table instances (see Chapter 3), features cannot overlap nor exist at the same time. Chapter 7 covers how to avoid problems with If loops in some detail, particularly how to suppress features manually and automatically to avoid overlap and coexistence. This information is very important to Pro/PROGRAM use generally. See the "Multiple Switches within the Assembly" section in Chapter 7.

Special Consideration for Interchange Groups

Type in a new part name while using family tables, and Pro/ENGINEER will automatically retrieve its correct instance. With interchange groups, on the other hand, Pro/ENGINEER only performs component interchange when the interchange group is either in memory, in the working directory, or in the search paths set up in the configuration file.

Five Fan Diameters and Five Different Motors

To make the five fan diameters available to the fan assembly, assign variables to *alt_blades.prt* and use Execute statements to pass them to the component. The assignment of variables takes place in the Relations section of the assembly file. The Relations section of *alt_box_fan.asm* showing assignment of variables for *alt.blades.prt* follows.

```
RELATIONS

IF WHICH_BLADE == 0

    FAN_DIA = 500

    BLADES = 5

ENDIF

IF WHICH_BLADE == 1

    FAN_DIA = 550

    BLADES = 5

ENDIF

IF WHICH_BLADE == 2

    FAN_DIA = 650

    BLADES = 7

ENDIF

IF WHICH_BLADE == 3

    FAN_DIA = 700

    BLADES = 7

ENDIF

IF WHICH_BLADE == 4

    FAN_DIA = 750
```

```
     BLADES = 7
ENDIF
```

➙ **NOTE:** *Variable assignment can occur as a user input statement as well as within an assembly's Relations section.*

In contrast, use a family table to build the five fan motors. A "generic" part for the table is assembled into *alt_blade_motor.asm* and table members recalled via the instance name. Instance names are assigned to the MOTOR_NAME variable. Type in the name, and the string stored in the variable passes down to the *alt_blade_motor.asm* assembly. A simple edit of the assembly's Add statement prompts retrieval of the specified family table instance. The Add statement controlling the family table instance to be used follows:

```
ADD PART (MOTOR_NAME)
INTERNAL COMPONENT ID 18
END ADD
```

The parentheses around the variable name tell Pro/ENGINEER to retrieve a family table instance instead of the generic part.

Variable choice and assignment likewise determine Pro/ENGINEER's interpretation of other family table parts such as motor screws. Again, enclosing the variable name in parentheses after the ADD statement enables Pro/ENGINEER to correctly interpret the variable as a family table instance.

```
ADD PART (MOTOR_SCREW_NAME)
INTERNAL COMPONENT ID 161
PARENTS = 147(#24)
END ADD


ADD PART (MOTOR_SCREW_NAME)
INTERNAL COMPONENT ID 162
PARENTS = 147(#24)
END ADD


ADD PART (MOTOR_SCREW_NAME)
INTERNAL COMPONENT ID 163
PARENTS = 148(#25)
END ADD


ADD PART (MOTOR_SCREW_NAME)
INTERNAL COMPONENT ID 164
PARENTS = 148(#25)
END ADD
```

Alternate Methods for Controlling Geometry Creation

To keep top level assemblies simple, consider feature workarounds to omit parameters needed by the lower level parts and subassemblies. The *alt_box_fan_housing.prt* part depends on motor and blade values to properly regenerate its depth. As shown in Chapter 7, these values can pass from the top level assembly via Execute statements. Alternatively, the depth itself can be defined in assembly mode as equal to the depth of *alt_blade_motor.asm* plus 30 mm for clearance. In this case, the motor and blade values need not pass from the top level for regeneration.

☞ **WARNING:** *Choose feature references carefully. Avoid creating circular references.*

Alternate Methods for Computing Costs

Chapter 7 provided an introduction to cost computing. Costs were computed by either storing "fixed" values as cost parameters within parts, or by calculating "rough" volumes using equations such as "length x width x height." Another method for computing cost is to manually instruct Pro/ENGINEER to calculate mass properties of the part or assembly and regenerate to update the cost parameters. Mass properties are accessed using Info | Mass Props.

Both the *alt_blades.prt* and *alt_box_fan_housing.prt* parts contain relations that use mass property volume (*mp_volume*). The relations section of *alt_blades.prt* showing cost calculation using mass property volume follows. The Relations section of *alt_blades.prt* showing cost calculation using mass property volume follows:

```
RELATIONS


/* This section calculates the cost of the component
/* by mutiplying the calculated volume by a known cost.


cost=mp_volume("")*.000025


END RELATIONS
```

Because mass property updates occur only at the user's request, the cost per part stays the same even if regeneration changes the part's overall mass or volume.

☞ **WARNING:** *Exercise caution when using mass properties to calculate other parameters such as cost. A statement saying the mass property has changed and must be recalculated is the only prompt to recalculate given during regeneration. In large assemblies, this prompt can fly by so fast you may not even see it. Be careful!*

✓ **TIP:** *You can force Pro/ENGINEER to automatically recalculate the mass properties by including the part or assembly names in the Massprop section of the Pro/program. The Massprop section would then resemble the following:*

```
MASSPROP

        PART                    NAME

        ASSEMBLY                NAME

END MASSPROP
```

Summary

This chapter focused on creating a top level assembly with an embedded Pro/program which also utilizes family tables and interchange groups. In particular, the chapter covered changes made to the components and dimensioning schemes compared to those used in Chapter 7. The techniques presented in Chapter 7 are not preferable to the techniques in Chapter 8, and vice versa. Both sets of techniques were discussed to demonstrate the range of conditions that Pro/PROGRAM can handle.

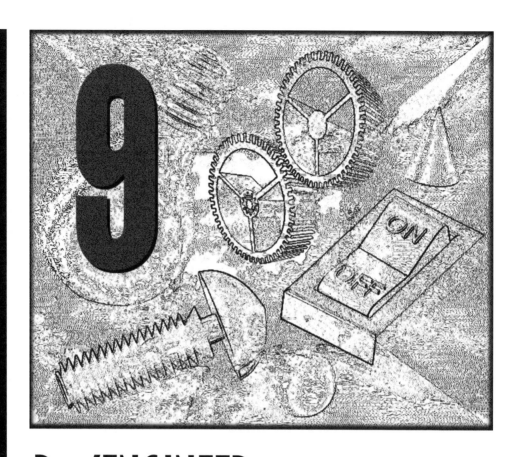

Pro/ENGINEER
Data Transfer

Round fan example used in this chapter.

➤ **NOTE:** *The material in this chapter is provided on the companion CD in the* fan1.doc *file. Another file,* fan2.doc, *includes process sheets for the round fan described in earlier chapters. Both* fan1 *and* fan2 *are Microsoft Word 6.0 files.*

The Advantages of Data Transfer

As demonstrated in earlier chapters, Pro/ENGINEER data enables product designers and developers to make rapid design changes. Not surprisingly, the same data can also make life easier and more productive for other company divisions. For example, given access to the design process, marketing can develop up-to-the-minute presentation material showing preliminary, detailed, and additional product features. Similarly, providing the finance department with data for product component dimensions enables its staff to make accurate, on-the-fly product cost projections. Pro/ENGINEER data can also be applied to documenting changes in cost, scheduling, and production and design activity, as well as to creating product development plans and spreadsheets.

Pro/ENGINEER data for PC applications.

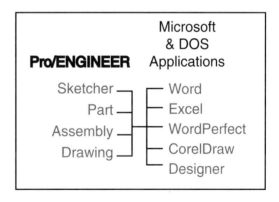

Typical network.

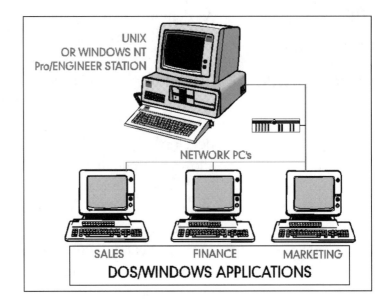

Developing such an environment, however, is not as simple. Pro/ENGINEER runs on high-end (translate to expensive) workstations. Company divisions whose work is geometrically non-intensive tend to use low-end (translate to less expensive) PCs, that is, machines typically not capable of running Pro/ENGINEER. The solution? Translate information from the Pro/ENGINEER platform to the PC platform, save money, and increase overall productivity.

✦ ***NOTE 1:*** *Windows NT and Windows 95 have made it possible to use Pro/ENGINEER on low-end workstations; however, the tips and techniques outlined in this chapter assume the more common practice of using Pro/ENGINEER on a high-end workstation.*

⊷ **NOTE 2:** *Pro/ENGINEER Rev. 17 supports OLE 2.0 linking with Windows based systems. These direct links are supported for parts, assemblies, and drawings.*

Cost comparisons between computer platforms.

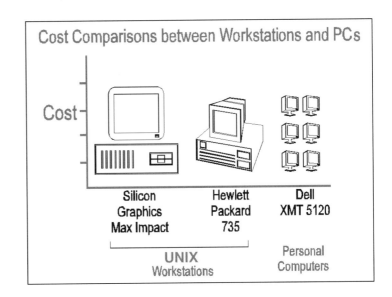

Cross-platform Configuration

Software

With the right configuration, Pro/ENGINEER can seamlessly transfer files from high-end CAD systems to PCs. Using the latest development information, Pro/ENGINEER can even automatically update illustrations, bills of material, and other product design data. In this chapter you will find methods, tips, and techniques for translating Pro/ENGINEER data into WordPerfect, Excel, CorelDRAW, Microsoft Word,

Vislab, and Visual Reality 2.0, but these are by no means the only Pro/ENGINEER-compatible applications.

Hardware

Pro/ENGINEER data files can take a big bite out of storage space (the fan files in this chapter require 200 Kb of disk space), so adequate storage is a must. Unless they store a number of other space-hungry applications, standard 600 Kb and 1 Gb PC hard drives should do the job.

Pro/ENGINEER data also tend to be RAM- and processor-intensive, particularly for a PC. Translation times are very good for a PC with 32 Mb of RAM, yet still reasonable for a PC with 8 Mb of RAM. An Intel 486 33-Mhz processor will work with Pro/ENGINEER data, but slowly; 90- and 120-Mhz Intel Pentium processors are, of course, faster. Graphics accelerators and advanced display drivers can also help speed things along, but the best investment for improving image regeneration on a slower PC may be to increase its available RAM.

Pro/ENGINEER generated high-resolution images (TIF and GIF data files) display best on monitors with at least a 1024 x 768 pixel screen. Increasing video RAM can improve color and refresh rates for these and other images.

Accomplishing Data Transfer

Data transfer is the procedure that makes it possible to copy plot files, photorealistic images, and bills of material from a Pro/ENGINEER working environment to a PC operating system. In short, data transfer is the means by which Pro/ENGINEER data is made useful to company divisions other than design and engineering.

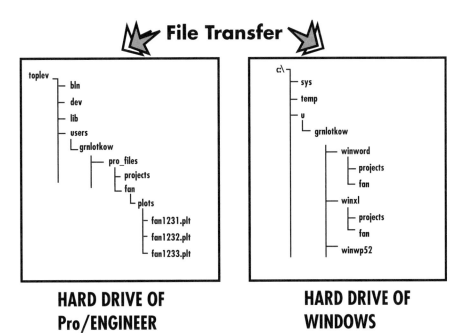

File transfer.

WFTP

Data transfer methods are almost as varied as the machines they serve. In Windows, the standard is the

Windows version of FTP (file transfer protocol), WFTP.

WFTP.

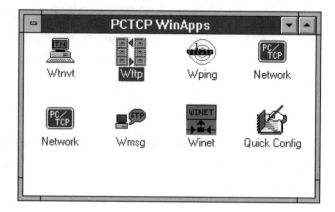

WFTP establishes a working relationship between the Pro/ENGINEER workstation and the PC user across a LAN (local area network). WFTP makes several modes of data transfer available, including ASCII, binary, and local 8.

Transfer mode	Data type
ASCII	Text files (bills of material)
Binary	Binary files (.gif/.tif)
Local 8	Files that use an 8-bit word size (not used)

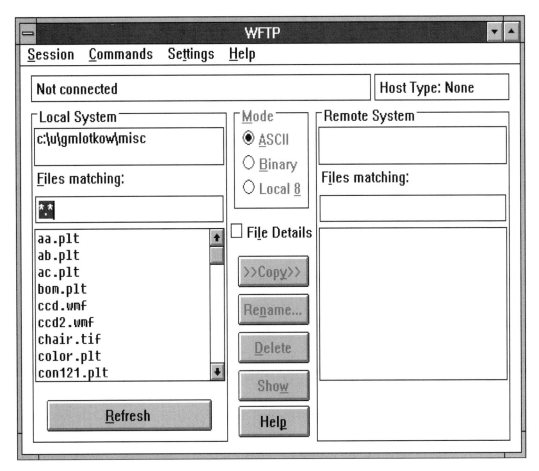

WFTP application.

To use WFTP to transfer Pro/ENGINEER plot (PLT) files between a remote workstation and a PC via a LAN, take the following steps:

1. Launch the WFTP program from the PCTCP WinApps application.

2. Connect to the remote workstation by selecting Session | New.

3. Type in the Host name or address, Username, and Password. The host name is the name of the remote workstation. The user logon and password are the name and password usually entered by users for access to the host machine. (If the logon or password is incorrect, permission to access files will be denied.)

Logon screen for WFTP.

4. Choose Connect. The New session window appears.

5. In the left column, select the directory tree on the local system or PC to which to copy files.

6. In the right column, select the files to transfer from the remote system.

7. Pick Copy to transfer the selected files.

When WFTP is first launched it displays the command line, "Not connected," and lists local system files on the left. On connection, it adds a list of remote system files to the display on the right and a selection of transfer modes including ASCII, Binary, and Local 8 in the center.

✓ *TIP:* *The best transfer mode for plot files is ASCII.*

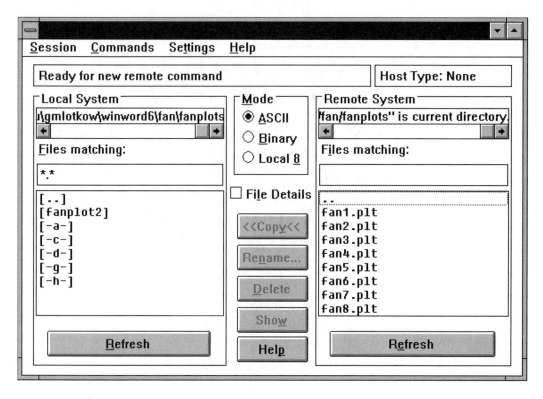

WFTP transfer.

During file transfer the command line changes to "Presently transferring" to indicate that the data transfer is in progress. When complete, the transferred data files appear on the left side, or local PC file list, and the files can be accessed for use with PC applications.

Data files can also be transferred via diskette, provided both the Pro/ENGINEER workstation and the PC have 3-1/4″ floppy drives. A file transfer utility can be used to copy files from the workstation to the diskette and from the diskette to the PC. Diskette file size limitations and the large size of Pro/ENGINEER files can make this option more time-consuming. For the same reason, diskette transfer is not efficient for transferring large graphics files or several files at once.

Format types.

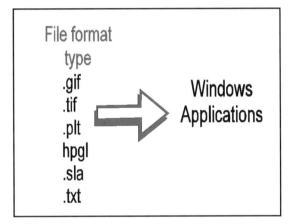

.plt Format

The *.plt* format is probably the most useful file format for exporting data from Pro/ENGINEER to Windows applications. Most Microsoft graphics applications can edit the *.plt* format, and *.plt* files also make for rapid plot file naming. This last feature is particularly beneficial given that Pro/ENGINEER creates multiple plot files when views are shown across multiple sheets. The *.plt* format is also the best for use in black and white wireframe illustrations.

Example of an HPGL2 plot file.

.tif and *.gif* Formats

Windows applications that incorporate color and create photorealistic images, such as those used to generate brochures and illustrations, work best with data exported in either the *.tif* or *.gif* format. *.Tif* and *.gif* files are also easily manipulated by photo and image editing applications such as CorelPAINT and Photo-MAGIC. However, *.tif* and *.gif* files do not tend to be compact, and can have strong negative effects on PC performance, particularly display speed. The best solution to this problem, as noted earlier, is to boost resident video and RAM.

Example of a .tif file.

.stl Format

Special rendering applications, such Visual Reality 2.0, import three-dimensional data or data exported using the *.stl* file format. With the use of *.stl* files, these applications apply special effects to create outstandingly photorealistic images. They also make interesting visual modifications to two-dimensional data using techniques such as texture mapping to

enhance the appearance of Pro/ENGINEER-gener-
ated surfaces.

Example of Visual Reality 2.0.

Exporting Data

Drawing Mode

Pro/ENGINEER can export data files from the Part,
Assembly, Drawing, Sketcher, and other modes, but
the Drawing mode seems to work best. Drawing
mode features that are particularly helpful when
exporting files include the following:

❖ Recording all changes to parts and assemblies

❖ Allowing for rapid generation of data in multiple-sheet drawings

❖ Providing parametric exchange between all Pro/ENGINEER modules

❖ Bringing all related part and assembly information into session once the drawing is retrieved

❖ Excluding unwanted data when generating plot files

Drawing mode can generate, assign a name to, and sequentially order separate plot files for each sheet of a multiple-sheet drawing.

To perform this operation, select Interface | Export | Plotter | Options from the Pro/ENGINEER menu. Set the option to All in the Current/All/Range field. The program names all sheets separately and assigns each a file number. The file number appears as the last character in the plot file name.

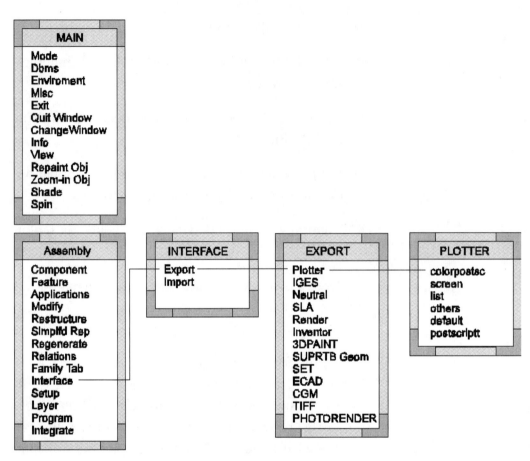

Interface / Export operations from Pro/ENGINEER.

The following example shows plot files made from a drawing with three sheets. The drawing name is *fan1.drw.*

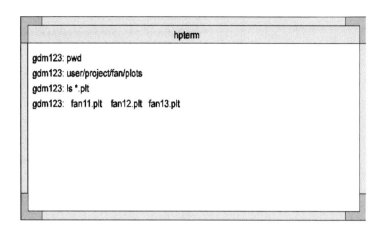

Series of plot files for the drawing fan1.drw.

➻ **NOTE:** *Remember to name files using five characters or less to ensure that numbering sequences are not lost during file transfer.*

The following table presents the export plot options available during the drawing mode export operation.

Plotter Options	Results
with format/without format	Removes drawing format from drawing file
current/all/range	Allows the user to specify the range of pages for the plot files
postscript	Will not translate, incompatible file format

Removing the drawing format from Sketcher data reduces file size and demands on resident RAM, making the data more suitable for illustration purposes in Microsoft applications.

Bills of Material

Bills of material exported from Pro/ENGINEER's Assembly mode are easily imported into PC word processor applications. To generate a bill of material file, select the sequence Info | BOM shown in the following illustration.

Generating a bill of material.

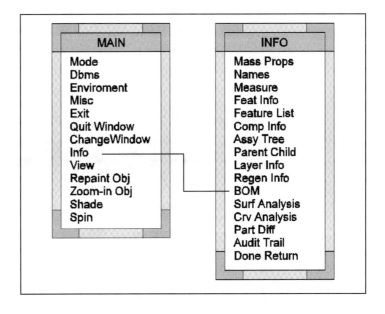

The next illustration shows a description for a bill of material generated by Pro/ENGINEER in UNIX.

UNIX description of a bill of material.

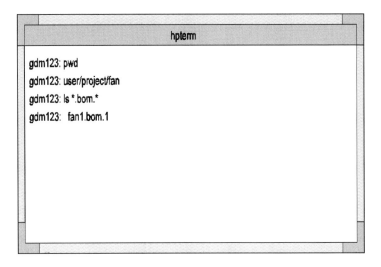

```
                          hpterm

gdm123: pwd
gdm123: user/project/fan
gdm123: ls *.bom.*
gdm123:  fan1.bom.1
```

A PC word processor translator can recognize the file type but may rename or not accept the UNIX name during translation. For example, Windows applications will not recognize the number of periods in the file name. To avoid confusion, rename the file using a text format suffix (*.txt*). A file name such as *fan1.txt* would be suitable and accurate because bills of material are essentially nothing more than text.

*Renaming a
bill of material.*

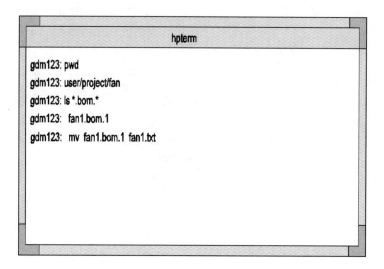

```
                              hpterm

gdm123: pwd
gdm123: user/project/fan
gdm123: ls *.bom.*
gdm123:   fan1.bom.1
gdm123:   mv fan1.bom.1 fan1.txt
```

Printers and Plotter Translators

Several HPGL2 printers support the export operation. The DesignJet printer is used to obtain illustrations with modifiable curves. Trial and error often guide discovery of an acceptable translator for various applications. The following table is a short summary of known plotter translators.

Plotter type	Results
Designjet (HPGL2)	Acceptable for Word 6.0, CorelDRAW, Designer, WordPerfect
hplaserjet4	Not favored; additional entities are often added to pictures
postscript	Will not translate; incompatible file format

Microsoft Translators

Microsoft application translators import and export files of various formats and make data interchange-ability between software applications possible. Translators are often installed on local or network PCs when the software is first installed. Prompts signal the user to select the translator that best fits the work requirements. Translators can also be added after the software has been installed.

File Maintenance

File maintenance is important to handling design information, and directories should be developed and used to keep data in order. Managing data for use with advanced features is even more crucial to ensuring that the software application can find and maintain data links.

A typical PC directory structure.

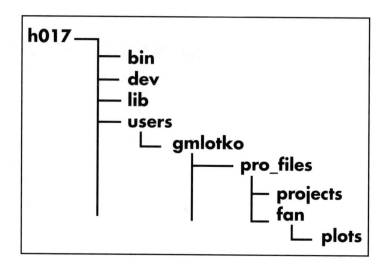

⚬ **NOTE:** *Using the network server's hard drive to store graphic and text files for network use ensures that all users have access to the same updated data, and prevents inadvertent use of outdated or inconsistent files.*

Pro/ENGINEER and Windows Applications

Several Windows applications can efficiently use Pro/ENGINEER data. As mentioned previously, CorelDRAW, Designer, Word, and WordPerfect are among proven applications. Later versions of the applications mentioned here have not been tested for the same functionality, but more than likely allow the same ease of use. The following table contains a summary of useful applications.

Software Name	Application	Examples
CorelDRAW	Graphics	Edit Pro/ENGINEER plot files; produce photorealistic artwork from *.gif* and *.tif* formats; autoreduce spline points on curves
Corel PHOTO-PAINT	Photo editing	Retouch photorealistic images (.gif/.tif formats)
CorelSHOW	Graphics	Perform presentations, slide shows
Designer	Graphics	Edit Pro/ENGINEER plot files; produce photorealistic artwork from *.gif* and *.tif* formats; use advanced features to combine artwork
Word	Wordprocessor	Edit *.txt* files; import *.plt*, *.gif*, and *.tif* formats
HiJaak Pro	Utility	View, edit, and process an image; convert one file format to another; organize graphic files
Visual Reality 2.0	Rendering	Enhanced texture mapping, morphing, special visual effects

Microsoft Applications

Microsoft Office

Microsoft Office usefully combines several software programs used in corporate offices, including word processor, spreadsheet, and database functions. It provides Pro/ENGINEER data translators and can create data links to other file formats for updating.

Microsoft Office applications.

Microsoft Word

Microsoft Word 6.0 is a word processing application that includes easy text and graphics formatting. It provides a variety of templates, or forms, for quick document creation. Word 6.0 drawing and text features complement Pro/ENGINEER for generating artwork and graphics. To import Pro/ENGINEER data into Word 6.0, open or create a new document, then choose Insert | Picture.

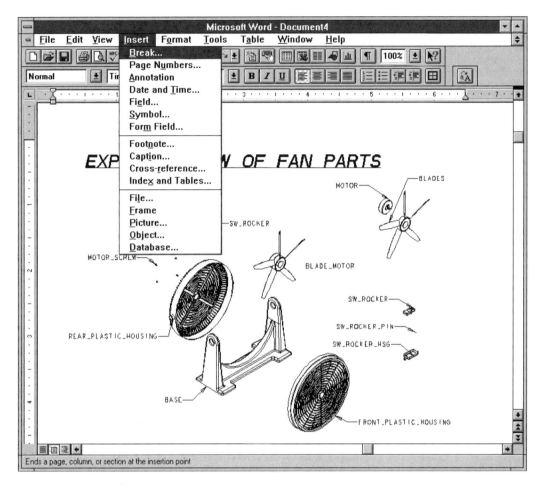

Word 6.0 Insert | Picture.

Locate the directory in which you have data stored and, from the List Files of Type field, choose the HP Plotter Print File (*.plt*). Choose the file you want to import into the document. The data file undergoes translation and appears in the document.

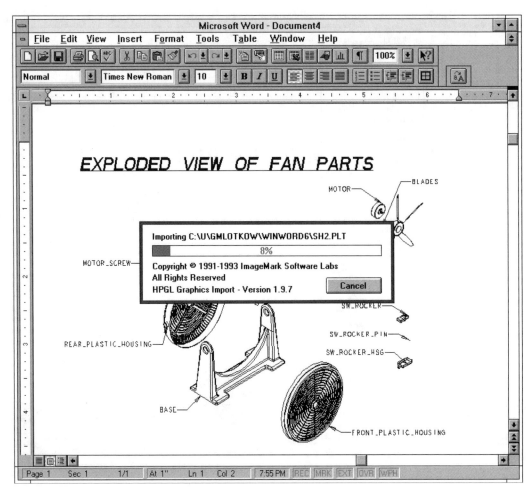

Word 6.0 importing a data file.

Using the mouse and clicking the picture once enables resizing. Double-clicking enables complete editing using Word's drawing tools.

Microsoft Excel

Microsoft Excel is a spreadsheet program which organizes data and performs automatic calculations in spreadsheets. Excel can produce charts and slide shows, summarize data, perform what-if scenarios, and is an excellent problem solver. It imports Pro/ENGINEER data in a manner similar to Word 6.0.

CorelDRAW

CorelDRAW 5.0.

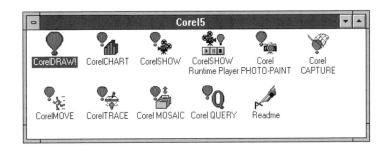

Two-dimensional wireframe data are modifiable when imported into graphics programs such as CorelDRAW 5.0 and Designer 4.0. To import files into CorelDRAW 5.0, go to the File menu, select Import, and choose the *.plt* file you would like to modify. A menu appears describing several HPGL options. These options allow you to adjust the pen width and color. Click OK to start the data translation. Choosing Edit | Select All, and then Arrange | Ungroup "ungroups" all the entities in the drawing, allowing spline modification. Use the Node Edit tool to stretch, bend, and add additional spline points.

*Example modifications
to switch rocker.*

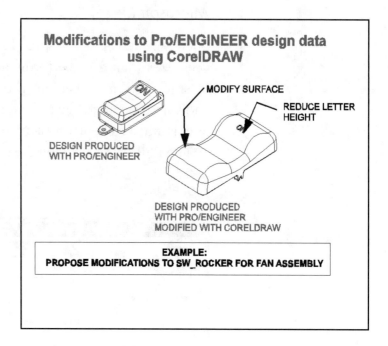

*Spline point
modifications.*

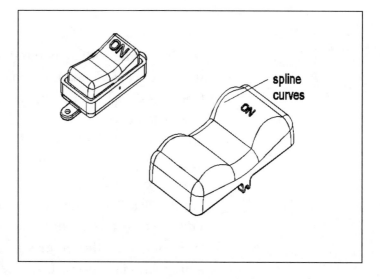

Visual Reality

Visual Reality 2.0 is a high-end PC tool that creates and renders object views. It produces outstanding effects within Pro/ENGINEER's stereolithographical format. Among other tasks, it can position objects in 3D space, illuminate them with multiple lights, and display them relative to "camera" views.

Visual Reality 2.0 applications.

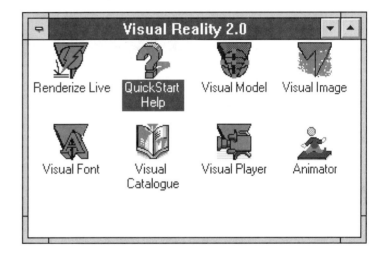

HiJaak PRO

HiJaak PRO is a very useful utility for the Windows working environment. It can view, edit, and process images. It can convert a file from one format to another, capture screens, and organize graphic files. Its image editing capabilities include cut and paste, cropping, changing size, rotating, modifying colors, color reduction, and other modifications.

HiJaak applications.

Object Linking

An advanced Microsoft Windows feature is linking files for automatic updating, known as OLE (object linking and embedding).

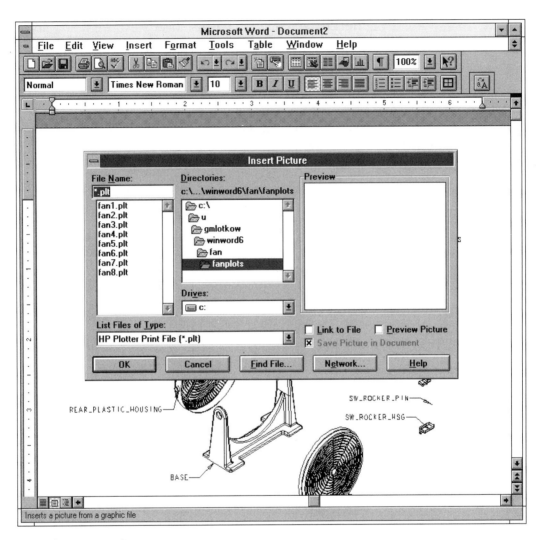

Sample process sheets.

Providing the link to a graphic file requires an additional selection when importing. Choose Link to File in the Insert | Picture menu to link to the selected graphic file. As the design process continues, the updated graphic files are transferred from

the Pro/ENGINEER workstation to the PC. To auto-
matically update the graphic links, choose Edit |
Links | Update Now.

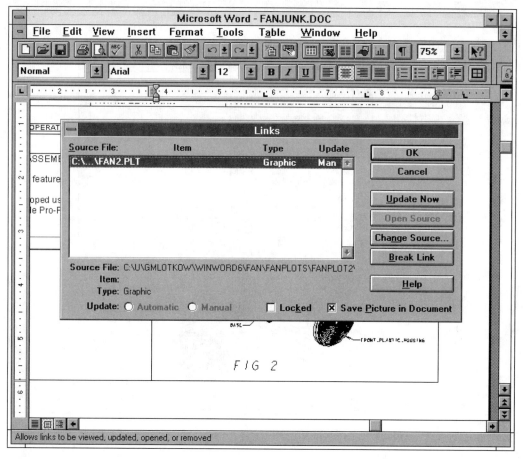

Update links.

This process is common to most Microsoft applications. Additional instructions are included in the
document *fan2.doc* stored on the companion CD.

Summary

This chapter outlined techniques to share information between high-end CAD systems and standard PCs. These techniques permit greater teamwork between development, marketing, finance, and other areas of your company. Improved communication at the front end in the development process will allow for streamlined development of new products. Hopefully this and previous chapters will allow you to achieve those results.

Molds, Analysis
and Rapid
Prototyping

Once the Pro/program is complete, designs are only partially complete. The previous chapters focused on Pro/program creation which allows rapid configuration and geometry changes. The program also reports estimated costs, allows updating of process sheets with the latest geometry in Microsoft Word, and creates the packaging required to ship the product. However, the following three steps remain:

1. Create the mold to build the parts (plastic) or stamp the parts (steel).

2. Mesh and analyze the parts (finite element analysis) to verify that they will meet design intent.

3. Create SLA mesh which results in an *.stl* file to enable rapid prototyping.

Due to the parametric nature of Pro/ENGINEER, the parts created to enable the three steps above (molds, finite element (FE) meshes and *.stl* files) will all update when changes are made to the base parts via the Pro/program. This chapter covers the creation of a parametric mold, setting up a mesh with local controls to allow rapid updates, and tips and techniques on the settings to use for generating a sound *.stl* file for rapid prototyping. These final steps allow you to truly automate Pro/ENGINEER. Any changes made via the Pro/program will cascade down to the molds required to create the geometry, update the mesh, and allow you to quickly create files for rapid prototyping.

The example used in this chapter is *sw_buttons_hsg_base.prt* provided on the companion CD. Because the part is a plastic component, Pro/MOLD is used to create the injection mold.

Creating the Mold

Once the parts are built, you can proceed to manufacturing analysis. The steps in the process follow:

1. Create a mold assembly (i.e., create mold solid parts and assemblies, assign shrinkages, create runners, etc.).

2. Check to determine whether the draft angle on the parts allows the mold to open.

3. Create split surfaces and partition the mold into independent "volumes."

4. Generate a solid representation of what will come out of the mold.

5. Analytically calculate whether the mold will fill properly.

While this section is not dedicated to performing the mold filling analysis, the latter is one of many that can be performed on a part. The next section in the chapter focuses on some of the typical analyses that could be performed on parts.

Creating a Mold Assembly

The first step is to create a solid rectangular part that represents the mold size (*mold_workpiece.prt*). The part is an ordinary Pro/ENGINEER part, that is, large enough to contain the number of parts to be created with each mold shot in the injection molding machine. In Pro/MOLD you also have the ability to store standard mold sizes (if your manufacturing facility has predefined mold base sizes) to make loading and unloading of the injection molding machines easier.

Simple mold part.

Select Mold | Create and begin to create the mold assembly. Pro/ENGINEER creates two files at this step. In this example, the files are called *sw_button_base_mold_asm.mfg* and *sw_button_ base_mold_asm.asm*. The *.mfg* file contains the manufacturing steps to be created later such as the mold opening distance. The *.asm* file is an ordinary Pro/

ENGINEER assembly that can be retrieved and worked on in assembly mode for reference, and so forth.

The objective is to construct a simple two-piece mold that does not require internal slides to operate. Thus, after creating the mold assembly file, the components must be placed in the assembly. Assemble *mold_workpiece.prt*, which represents the mold geometry.

Add two copies of the *sw_button_hsg_base.prt*. As you add each copy of *sw_button_hsg_base.prt*, Pro/ENGINEER automatically creates "reference copies" of each part. The copies are parametrically linked to the original part, but can be modified further (e.g., additional draft angles and shrinkage factors).

New names for the reference geometry files must be assigned at this time. In the example they are called *sw_button_hsg_base_ref.prt* and *sw_button_hsg_base_ref2.prt*.

Shrinkage must be considered for plastic components when designing molds. Shinkages can be assigned in Pro/MOLD by scaling or dimension. Use the By Scaling option which globally assigns shrinkage factors in the x, y, and z directions independently.

You can also select the coordinate system that you want Pro/ENGINEER to use when performing the calculations. The shrinkage factors are independent, and as such you can assign x only, y only, z only or any combination of the factors. In the example model, a uniform factor was assigned to each of the reference parts: *sw_button_hsg_base_ref.prt* was given a 5% shrinkage factor in X, Y and Z, while *sw_button_hsg_base_ref2.prt* was assigned a 15% shrinkage factor. The difference in shrinkage factors is readily apparent in the relative sizes of the two parts in the assembly. (Pro/ENGINEER "scales up" the geometry to account for shrinkage.)

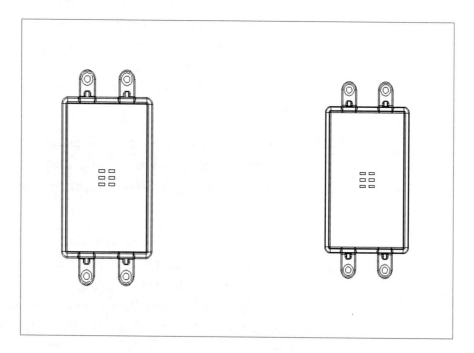

Two-cavity mold with different shrinkage rates applied to each cavity.

The By Dimension option allows you to select individual part dimensions and assign shrinkage factors. In a large model this can be tedious, and most shrinkage factors are not created in this way. The By Dimension option is generally used in conjunction with the By Scaling option to control individual dimensions which, due to peculiar geometrical design, will cause excessive shrinkage compared to the overall part. The By Dimension values entered will override any values assigned using the By Scaling option.

For instance, assume that you are working with a large box with a few very tall and thin fins. In this case, you might first assign an overall By Scaling shrinkage factor of 5%. However, based on prior experience with this type of part you might know that the height of the fin generally shrinks by 10%. You could then use By Dimension to assign a 10% factor to the height of the fin only, leaving the rest of the model at the standard 5% rate. This will not cause the height of the fin to shrink by 15%; Pro/ENGINEER will instead merely override the 5% value for this single dimension to be 10%.

Check Draft Angle to Permit Mold Opening

After the mold assembly is defined, the next step is to begin to figure out where the parting surface(s) should be. To aid in this selection Pro/MOLD allows you to quickly define a "pull" direction for the mold and analyzes whether the individual "reference parts" have any draft problems that might lock the mold when in operation. To perform the check, select Mold Model | Mold Info | Draft Check. You will then be queried to respond to the following:

❖ Select a datum plane or surface perpendicular to the pull direction.

❖ Select the reference part that you wish to select.

❖ Select the pull direction.

❖ Define the minimum required draft angle.

Pro/ENGINEER then calculates the draft on all part surfaces and reports back with color codes on each surface. All surfaces that fall below the minimum draft are highlighted.

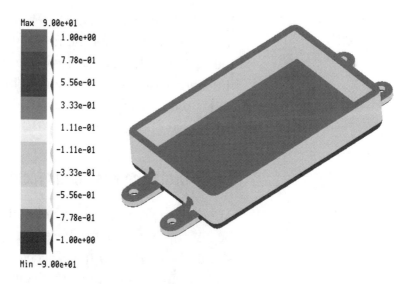

Output of draft check.

The above tool can help you narrow down where the parting line should be in the geometry which then allows you to proceed to the next step and create the split surfaces. If the part was properly designed, this should be a somewhat trivial task to verify where the designed parting line is on the geometry.

The housing part used in the Pro/MOLD has no draft. This obvious error was left in the file so that you can run the draft check and see that the edges with rounds easily pass the minimum draft check, while the vertical walls highlight as failing. At this stage, if this were a real mold, you would then be able to either modify the original part to include draft, or you could apply draft to the reference parts created from the original. Either method would create an acceptable part to mold.

Create Split Surfaces and Partition the Mold

The next step is to define the actual parting surfaces which will allow Pro/MOLD to "split" the solid mold into volumes that can then be extracted into individual mold pieces. Pro/ENGINEER's full functionality is available for creating/copying surfaces to use for the split surfaces. In the example surface copies of the two individual reference parts were used, and then additional surfaces created to "connect" the two models in the mold. The result was a single surface to be used to "split" the mold volume into two pieces.

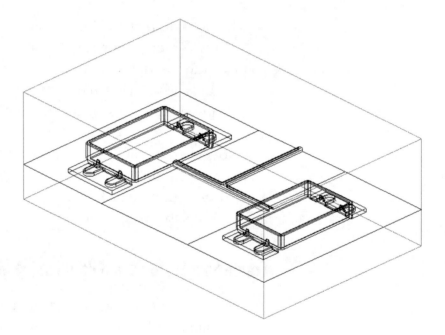

Mold with split surface.

If the mold had more than two pieces you would merely have to create additional surfaces that split the mold into additional individual volumes. When the surfaces are defined, the next step is to select Split and Pro/MOLD will then intersect all splitting surfaces with the solid mold volume and generate individual volumes defining the separate mold pieces. These individual volumes are basically assembly features at this point rather than independent "parts." To create the individual parts of the mold you must then use Extract Volumes which will then query you to provide a name for each of the individual volumes in the mold. In the example the parts and volumes were given the same names: *mold_top* and *mold_bottom*. The parts have the customary *.prt* extension.

✓ **TIP:** *Defining all surfaces that you will be using to create the split is recommended before Pro/MOLD generates the split.*

Go into the mold assembly. With the Mold Opening | Define Steps menu selection you can then assign the sequence and the distance that each piece of the mold will move. To view, use Explode.

Exploded view of two-cavity mold with reference parts.

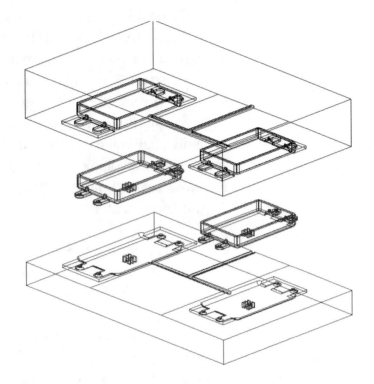

Generate a solid representation of what will come out of the mold ("molding")

Runners and sprues were included in the example. Pro/MOLD also contains the functionality to create a "molding"—a solid is generated showing you the entire contents of the mold, including runners and sprues. In the case of a multiple-cavity mold, this is a very useful capability as it allows you to then perform a volume calculation to determine the amount of wasted material in the runner system.

Exploded view of two-cavity mold with "molding."

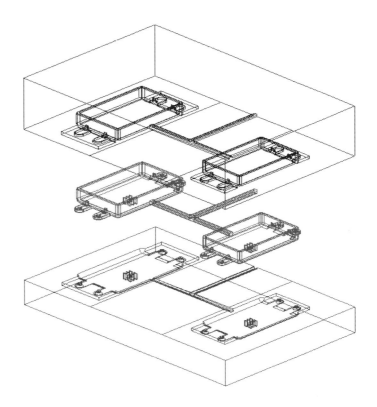

This section provided a quick walkthrough of the steps required to create a mold for a plastic component using Pro/MOLD. The associativity to the original part offers incredible time savings when changes to the parts force changes in the mold design.

Mesh and Analyze the Part

The Pro/program allows you to quickly develop multiple design scenarios but still leaves many questions about the integrity of the design. For instance, is the design strong enough to meet all engineering requirements for strength and durability? The design model is used in conjunction with computer aided engineering analytical methods to assist in up-front engineering decision making. Finite element analysis (FEA) is a method used to analytically predict the behavior of an electronic model. Boundary conditions, loads, and constraints are applied to a finite element mesh model to simulate real-life testing conditions. These analytical prototypes are based on assumptions, experience, and correlation to test data.

The common thread that exists among all analyses such as structural, thermal, kinematic, and CFD is that they all originate with a finite element mesh model (FEM). Systems and subsystems are the focus of analysis today so meshing needs to become faster and more reliable. Pro/ENGINEER solid models provide analysts with the ability to quickly generate a hand mapped, or auto mesh. Automeshing capability is the key to the future for finite element analysis to keep pace with the fast-paced design revolution. This section will primarily focus on tips and techniques for improving mesh and analytical results using Pro/ENGINEER.

Designing for FEA

Numerous key factors play a role in performing a successful finite element analysis within Pro/ENGINEER. Two of the most deciding factors related to the accuracy of the results of an analysis are related to the automesh quality and automesh quantity. Automesh quality topics deal with poorly shaped elements, inconsistent node spacing, high aspect ratios, and element warpage. Whether using solid or shell elements, 75 percent of most automeshing problems originate from a poorly constructed solid model. Automesh quantity topics deal with the number of elements needed to accurately define the stiffness of a design model. Stress and deflection results can be up to 100 percent off with a coarse mesh model compared to a fine mesh model.

Automesh Quality

Automesh quality is dependent on the topology of the design, the amount of information present when automeshing, and the quality of the math surface that defines the model. In most cases fine feature details need not be present when meshing the model. These features are typically small rounds and fillets, chamfers, short boundary edges, and small holes. Typically, only the areas of focus require fine feature details to be included in the mesh. The rest of the model is used to create elements for mass and stiffness.

The quality of the math surface is also important because the meshing algorithm is very sensitive to tangency and continuity blending between each surface patch edge. Pro/ENGINEER seems to start generating the elements in the middle of a surface patch and work outward toward each edge boundary. Sometimes compromises are made at multiple patch edges for shape and triangular elements. Checking every potential tricky area for patch connectivity is recommended.

Potential Problems

Potential problems resulting from a poorly automeshed model are listed below.

❖ *Increased mesh times.* Longer mesh times reflect too much detail in the model, resulting in increased CPU time, RAM, and swap space.

❖ *Large element counts.* Extraordinarily high element counts result in longer solver solution times thereby reducing analysis efficiency.

❖ *Small mesh elements.* A large number of small mesh elements results in an irregular pattern of mesh densities. This promotes the creation of bad transition elements from small to large mesh areas, high aspect ratios (higher than 10 to 1), and even collapsed elements. Depending on the solver of choice, small mesh elements can create major headaches.

Potential Solutions

Potential solutions to a poorly automeshed model follow.

❖ *Simplified model.* Defeature the model in areas of lower priority concern by removing fine feature details such as drafts, rounds, fillets, and chamfers.

Detailed model.

Defeatured model.

Model defeaturing is accomplished by practicing the sound design principles listed below.

❖ Putting detail features at the end of the model.

❖ Keeping track of and removing unwanted parent/ child relationships.

❖ Automatically assigning detail features to a layer for quick deletion or suppression.

❖ Creating family table instances of multiple analysis models.

❖ When working with thin-walled features, use a global shell command and solid thin features for automatic surface autopairing. Keep the manual pairing of surfaces to a minimum!

❖ *Geometry checks.* Geometry checks indicate the existence of entities (edges, faces) within the model that are small with respect to the global size of the model. This will also cause the mesh generator to spend a lot of time trying to distinguish the mesh pattern across the small edges. Quite often it creates an artificially high local element count or no mesh at all. Use Info | Measure | Min Radius to automatically highlight small areas within the model. Eliminate all geometry checks if possible! Return to the model and explicitly align corners or features in the sketcher to eliminate undesirable short edges or make reasonable modifications to the geometry.

❖ *Small surface patches (shells).* When surfacing, it is generally a good design practice to keep the number of surface patches to a minimum. Extra surface patches result in inconstant and high mesh densities. Surfacing from multiple boundary edges often generates unwanted patches. Use control points to adjust the flow lines between hard split points when using boundary blended surfaces. In addition, try creating a C2 approximate composite curve through the edges of the desired sections prior to the creation of the surface. The outcome is a one-piece boundary which produces improved mesh results.

Automesh Quantity

A model's element count drastically determines the outcome of the analysis results. A certain number of elements are needed to accurately define the stiffness of a model such as the number of elements through any single wall thickness. At least three elements through the wall thickness of a part are typically needed to accurately predict the proper bending stresses. Next, a certain number of nodes around key areas such as bolt holes and slots are required to accurately add constraints, loads, and boundary conditions.

Case Study

A case study of an FEA using three automesh densities is described in this section. The analysis performed was used to determine the deflection of the fan switch button housing as a 300 N load is applied at one edge. For this analysis the *sw_buttons_hsg_base.prt* was simplified (small rounds and details were removed). The geometry that was meshed and analyzed is provided on the companion CD in *sw_buttons_hsg_fea.prt*. The design criteria in the case study are to determine whether the housing deflects more than .5 mm near a proposed weight reducing slot.

Fan switch button housing.

Loads and boundary conditions.

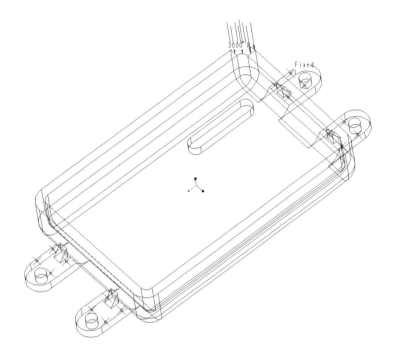

Mesh Density versus Deflection Results

A deflection analysis was performed on the fan switch housing using three drastically different automesh densities. The results ranged from a maximum deflection of .149 mm at an element count of 990 to .587 at an element count of 36717. Results from the analysis confirmed how important the mesh density is to the outcome of the analysis. The different meshed and result models appear below.

Mesh density1 (990 elements).

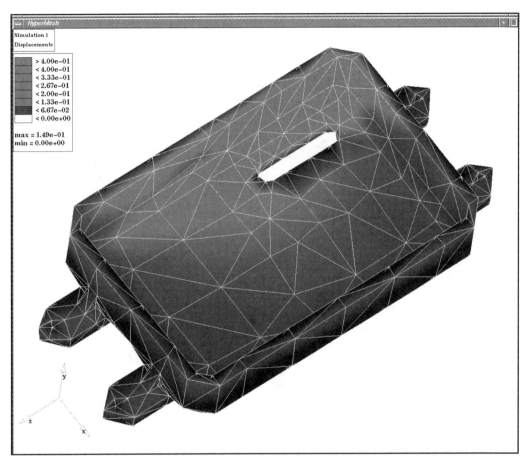

Deflection results 1 (.149 mm).

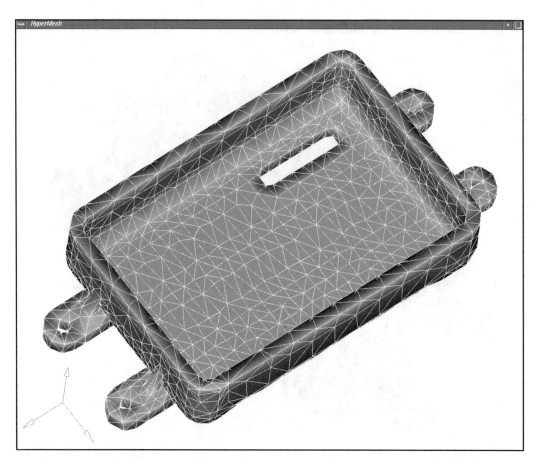

Mesh density 2 (5444 elements).

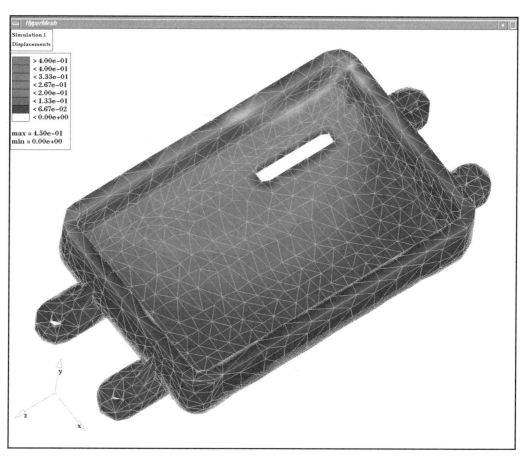

Deflection results 2 (.450 mm).

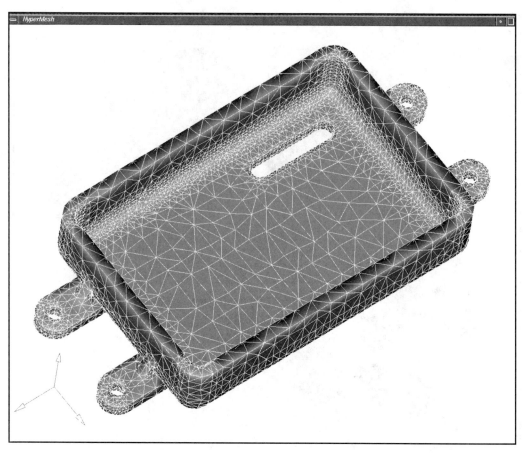

Mesh density 3 (36717 elements).

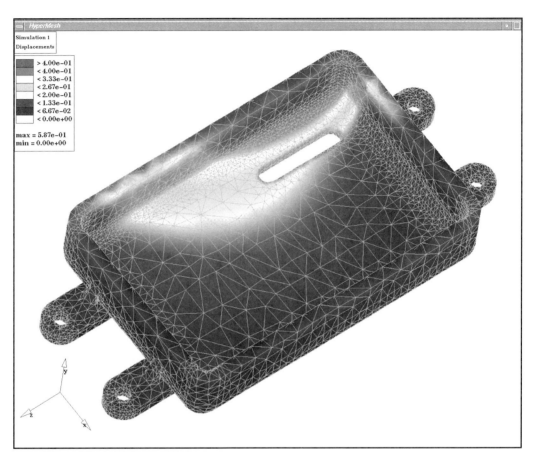

Deflection results 3 (.587mm).

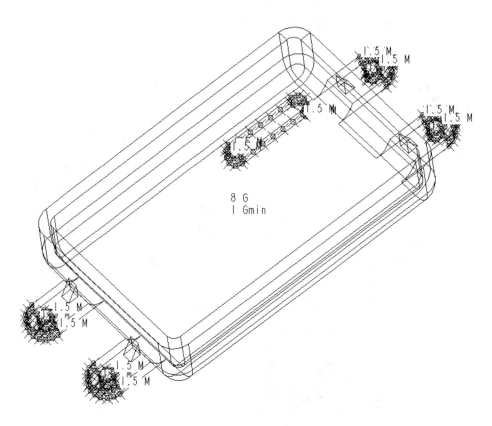

Mesh density versus deflection table.

Meshing Tips

❖ Mesh with Pro/ENGINEER default mesh controls initially to determine whether the model will even mesh. Fix geometry mesh problems at this time before tighter mesh controls are applied to the model. If you start off with very fine mesh control, it may take 45 minutes to detect geometry problems while you wait for Pro/ENGINEER to attempt to finish the mesh, whereas with the default conditions it may be exposed within minutes.

❖ Gradually add local mesh control to faces, edges, and hard points before global mesh controls are applied. Because Pro/ENGINEER only provides outside surface mesh control, this process will help you to determine the mesh density before adding the global inside mesh control.

❖ Add global Min and Max mesh control to fine tune the overall mesh of the part. Depending on the part size, introduce a global Min first and then a global Max to help the mesh hone in on an acceptable average element size. Pro/ENGINEER takes these inputs into consideration but does not exactly adhere to the hard number.

Local and global mesh control.

MESH	Elements	Nodes	Deflection
1	990	370	.149mm
2	5444	1801	.450mm
3	36717	9026	.587mm

❖ Use a boundary element mesh to identify problem areas on the outside surface. This process will allow the user to fix these areas which will result in a better solid element mesh.

❖ When the model absolutely will not mesh, start to section the model with cuts until the mesh is generated. Again this is a technique used in isolating problem areas within the solid model.

❖ Confirm that FEM_MAX_MEMORY in your *config.pro* file is set high. Depending on the size of the model, Pro/ENGINEER may sometimes require up to 750 Mb of space to mesh the models.

❖ Use as much RAM as possible!

❖ The automesher uses 90 percent of the memory needed for the first pass. Use hardware system analytical and graphical checks to monitor the amount of CPU, memory, and swap space used during the mesh generation. If a memory error occurs, adjust the global Max mesh control until the first pass threshold is broken. A memory error frequently occurs in the first pass. Multiple iterations may be necessary to generate a large mesh size if machine memory is limited.

SLA Meshing

This section focuses on how to create and refine an SLA (stereolithography apparatus) file, that is, the triangulation or mesh of the model's boundary. SLA is merely a format in which most solid modeling softwares and all stereolithography machines or rendering packages can communicate. Stereolithography is a method to produce rapid prototypes of components and assemblies that permit studies of manufacturability, assembly clearances, and assembly methods, among others, and the ability to see and hold your work.

Perhaps the most significant characteristic of stereolithography is that no money must be spent on tooling in order to complete the abovementioned studies, not to mention additional expenditure for modification and scrapped tooling following changes. Stereolithography requires that a mesh be as accurate as possible to convey part features. Likewise, rendering packages allow for similar studies to be performed without having to spend money on stereolithography components. Some of these rendering packages also allow for interference checks while attempting to assemble the component into position and getting the assembly tools into position for fastening.

Considering the importance of creating an accurate mesh, certain settings can be modified to derive the desired outcome. The SLA file is created by selecting Interface | Export | SLA from the MODE menu. The following two settings are changeable:

❖ *Chord Height.* This option changes the maximum distance between the chord and the actual surface. Therefore, the smaller the number, the tighter the mesh will stay to the original surface. This option will only modify the mesh if the component has curved surfaces. The lower bound is calculated by a function of the part's accuracy with the upper bound corresponding to the part size.

❖ *Angle Control.* This option, used mainly for small radius surfaces, calculates the extent of the additional change to the mesh. A small radius is defined by multiplying 10% to the part size. Any radius below this number will be modified. Therefore, if a value of one (1) is entered, there will be no improvement. The default for this number is .5 with a range from 0 to 1 with 0 being the highest refinement.

Having only two constraints can make it difficult to achieve high quality mesh for some components, especially if the components are long in one dimension and short in another. This situation makes it difficult to control how the mesh reacts on bosses and holes. These features tend to look like pyramids instead of cylinders. However, the techniques described below serve to further improve the mesh.

1. *Changing the component accuracy.* Decreasing the number in the accuracy variable increases the accuracy, and vice versa. The component accuracy will not be lowered enough for some components because of features not being able to regenerate. Consequently, this technique is not very reliable.

2. *Transfer the component through IGES.* You can create a new component with very low accuracy by using a translated file in which to read the IGES file. This technique is fairly reliable; however, because the new component is divorced from the original component, updates due to changes are not automatic.

3. *Creating a copied surface in a separate assembly.* Create an assembly and assemble the original component. Then create a new part while in the assembly. Select Component | Create | Part | Enter part name | Solid | Surface | Copy | Done | Surfaces | Surf and Bnd. Select a surface as the seed or beginning. Then select the same surface as the bounding surface to create a complete copy of all surfaces in the original component. Finally, create a solid from the surface quilt, and turn the part accuracy as low as necessary for the required mesh density. This technique is preferred due to associativity between the components. In order to update the copied surfaces after a change occurs, simply retrieve the assembly and Regenerate | Automatic.

The *sw_buttons_hsg_base* is used in the following to illustrate the above techniques. All graphics will be focused on the corner shown below.

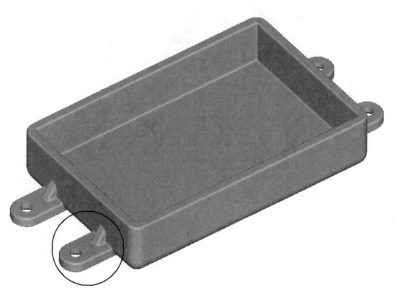

sw_buttons_hsg_base part with viewing zone located.

When the part is meshed with the default settings (Chord Height = 0.3578, AngleControl = 0.50), both the mounting tab and hole are polygons and the rounds look like a chamfer.

Mesh with default settings, Chord Height = .3578 and AngleControl = 0.50, resulting in 1532 triangles.

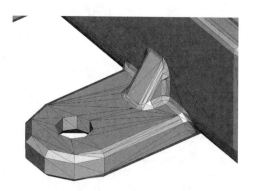

The resulting mesh density could be used for a rendering package, but not for a rapid prototype.

The mesh should be finer in order to better fill out the geometry. Thus, a smaller chord height would be appropriate. The method of determining the smallest chord height that Pro/ENGINEER will allow is to enter a zero (0) as the chord height parameter. Pro/ENGINEER will then prompt the acceptable range for the mesh. As shown below, the mesh is much closer to the original model upon entering the smallest chord height.

Mesh with Chord Height = .0108 and AngleControl = 0.50 resulting in 7524 triangles.

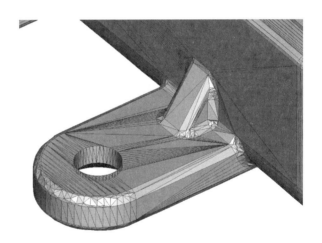

The only variable remaining is the angle control. As mentioned above, the smaller the number the greater the refinement. When the part is meshed again with the angle control set to zero (0), the mesh is even closer to the original surface.

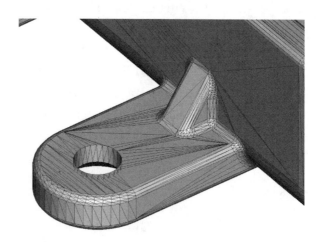

Mesh with Chord Height = .0108 and AngleControl = 0.00 resulting in 7048 triangles.

The above mesh should be sent out for a rapid prototype. This mesh could also be used in a rendering package; however, the number of triangles in the mesh is greater and could cause the rendering package to slow down. Hardware platforms rather than rendering packages may cause problems in terms of triangle numbers. Dynamically rotating a mesh requires extremely intensive calculations. Therefore, the greater the number of triangles in the rendering package, the slower the system will run.

With the use of technique 1 described above, the chord height can be as small as 0.0072 if you set the part accuracy from the default 0.0012 to 0.0008. Although the difference is a mere .0036, the mesh shows an improvement of 2690 triangles. The previous mesh contained only four rows of triangles across the rounds compared to five after using the first technique. The mesh is now more accurately represented.

Mesh with Chord Height = .0072 and AngleControl = 0.00 after changing the part accuracy to 0.0008 resulting in 9738 triangles.

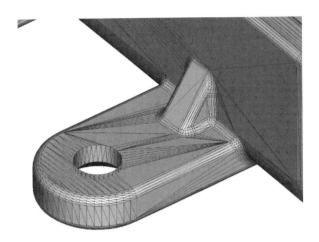

Export the surfaces of the *sw_buttons_hsg_base* part using the technique 2. Create a new part and import the surfaces. If you try to create an SLA mesh now, the smallest chord height possible is 0.0107. Although this action results in an improvement, the part accuracy must be increased to take advantage of having no history to replay. Set the accuracy to 0.0004. The chord height is now able to go as low as 0.0036. Mesh the part with this chord height and the angle control set to zero (0). The number of triangles nearly doubles, from 9738 to 18168.

Mesh with Chord Height = .0038 and AngleControl = 0.00 with the part accuracy set to 0.0004 resulting in 18168 triangles.

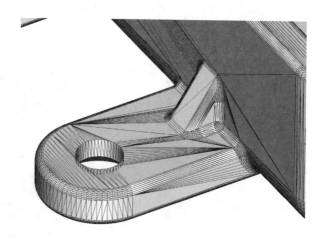

↝ **NOTE:** *Technique 2 does have a limitation. If your component has many "geom checks," quilting the imported surface at a low accuracy may be problematic. The only way to resolve this problem is to fix the component before you export the surfaces.*

Technique 3 is the preferred method. Create an assembly and assemble the *sw_buttons_hsg_base* part first. Creating the default datums and a coordinate system is not necessary. Next, while in the assembly create the surface copied part using the following menu picks: Component | Create | Part | Enter part name | Solid | Surface | Copy | Done | Surfaces | Surf and Bnd. Next, save the assembly and quit window out of the assembly. Retrieve the new part and create a protrusion using the quilt. At this time you can change the part accuracy to be as small as you wish. The major advantage of this technique is the ability to update the SLA mesh whenever the component changes by simply retrieving this assembly and picking Regenerate | Automatic. Set the accuracy to 0.0004 and create an SLA mesh. The smallest chord height possible is 0.0036. Results of setting the angle control to zero (0) appear in the next illustration.

Mesh with Chord Height = .0036 and AngleControl = 0.00, with the part accuracy set to 0.0004 resulting in 17432 triangles.

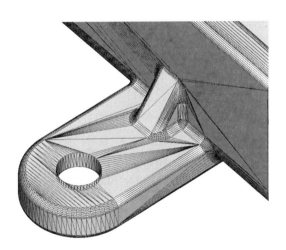

Summary

This chapter focused on the steps required to create a parametric mold from the model, set up local mesh controls to allow rapid mesh generation when part geometry changes, analyze the part to ensure that it meets design intent, and create *.stl* files to generate rapid prototype parts.

Index

G

generic parts and instances
   automatic interchange 38
   common "generic features" 46
   common "standard parts" 42
   non-repetition of common information 38
   part files management 39
   tabular drawing creation 42
geometry
   automatic rebuilding of 14
   controlling creation of 191
   controlling with relations 10, 114
   program creation for 10
.gif files
   display requirements 200
   in data transfer 219
graphics
   .plt editing capabilities 207
   file formats for 220
   file linking for 227

H

hard coding
   of values 70
   of variable 125
hard drive requirements
   for Pro/ENGINEER 200
hardware
   for Pro/ENGINEER data transfer 200
Hijaack Pro 219, 226

I

If loop
   Execute statement inclusion within 182
   for adding components 138
      137
   for foam and box parts 146
   in input section
      144
   use with features 186–187
   with Instantiate command 52
If statement
   use with Add Part statement 98
   use with End If statement 99
If...Else clauses 103, 104
If...Else...Endif statements

value assignment using 21
information
   non-repetition of 38
information window
   in Pro/program 33
input
   additions to area 67
   in top level assembly 154, 183
   use with variable 23
input statement
   assigning variables with 189
   consolidating subassembly levels 181
   error in 102
   in Pro/program 128, 132, 141
   passing of values in 101
   passing of with Execute statement 154
input variables
   number type 23
   string type 23
   yes-no string type 23
Insert mode
   compared to Interact statement 105
instance
   and features 46, 50–51
   choosing of 53
   FEA model as 52, 55
   generic 55
   modifying 45
   names for 189–190
   retrieving 40, 187
   use within family table 46, 50–51
instance accelerator files 40
Interact statements
   compared to Insert mode 105
interchange
   adding interchange assembly to 67
   and Pro/PROGRAM 65, 67
   automatic 38
   compared to assembly 59
   creating and using 2, 12, 61–62
   family table use with 179
   functional interchange 61, 62
   geometry and features restriction for 70
   simplify interchange 62
   special considerations for 187
   using 59
interchange groups
   and Pro/PDM 60
   creating 60–61
   indentifying with parentheses 185

More OnWord Press Titles

Computing/Business

Lotus Notes for Web Workgroups
$34.95

Mapping with Microsoft Office
$29.95

Geographic Information Systems (GIS)

GIS: A Visual Approach
$39.95

The GIS Book, 3E
$34.95

INSIDE MapInfo Professional
$49.95

*Raster Imagery in Geographic
Information Systems*
$59.95

INSIDE ArcView
$39.95 Includes CD-ROM

ArcView Exercise Book
$49.95 Includes CD-ROM

ArcView Developer's Guide
$49.95

*ArcView/Avenue Programmer's
Reference*
$49.95

101 ArcView/Avenue Scripts: The Disk
Disk $101.00

INSIDE ARC/INFO, Revised Edition
$59.95 Includes CD-ROM

ARC/INFO Quick Reference
$24.95

INSIDE ArcCAD
$39.95 Includes Disk

MicroStation

INSIDE MicroStation 95, 4E
$39.95 Includes Disk

MicroStation 95 Exercise Book
$39.95 Includes Disk
Optional Instructor's Guide $14.95

MicroStation 95 Quick Reference
$24.95

MicroStation 95 Productivity Book
$49.95

Adventures in MicroStation 3D
$49.95 Includes CD-ROM

MicroStation for AutoCAD Users, 2E
$34.95

INSIDE MicroStation 5X, 3E
$34.95 Includes Disk

MicroStation Exercise Book 5.X
$34.95 Includes Disk
Optional Instructor's Guide $14.95

MicroStation Reference Guide 5.X
$18.95

Build Cell for 5.X
Software $69.95

101 MDL Commands (5.X and 95)
Executable Disk $101.00
Source Disks (6) $259.95

Pro/ENGINEER and Pro/JR.

INSIDE Pro/ENGINEER, 2E
$49.95 Includes Disk

Pro/ENGINEER Exercise Book, 2E
$39.95 Includes Disk

Pro/ENGINEER Quick Reference, 2E
$24.95

Thinking Pro/ENGINEER
$49.95

Pro/ENGINEER Tips and Techniques
$59.95

INSIDE Pro/JR.
$49.95

Softdesk

INSIDE Softdesk Architectural
$49.95 Includes Disk

Softdesk Architecture 1 Certified Courseware
$34.95 Includes CD-ROM

Softdesk Architecture 2 Certified Courseware
$34.95 Includes CD-ROM

INSIDE Softdesk Civil
$49.95 Includes Disk

Softdesk Civil 1 Certified Courseware
$34.95 Includes CD-ROM

Softdesk Civil 2 Certified Courseware
$34.95 Includes CD-ROM

Interleaf

INSIDE Interleaf (v. 6)
$49.95 Includes Disk

Interleaf Quick Reference (v. 6)
$24.95

Interleaf Exercise Book (v. 5)
$39.95 Includes Disk

Interleaf Tips and Tricks (v. 5)
$49.95 Includes Disk

Adventurer's Guide to Interleaf LISP
$49.95 Includes Disk

Other CAD

Manager's Guide to Computer-Aided Engineering
$49.95

Fallingwater in 3D Studio
$39.95   Includes Disk

Windows NT

Windows NT for the Technical Professional
$39.95

SunSoft Solaris

SunSoft Solaris 2.* for Managers and Administrators
$34.95

SunSoft Solaris 2.* User's Guide
$29.95   Includes Disk

SunSoft Solaris 2.* Quick Reference
$18.95

Five Steps to SunSoft Solaris 2.*
$24.95   Includes Disk

SunSoft Solaris 2.* for Windows Users
$24.95

HP-UX

HP-UX User's Guide
$29.95   Includes Disk

HP-UX Quick Reference
$18.95

Five Steps to HP-UX
$24.95   Includes Disk

OnWord Press Distribution

End Users/User Groups/Corporate Sales

OnWord Press books are available worldwide to end users, user groups, and corporate accounts from local booksellers or from Softstore/CADNEWS Bookstore: call 1-800-CADNEWS (1-800-223-6397) or 505-474-5120; fax 505-474-5020; write to SoftStore, Inc., 2530 Camino Entrada, Santa Fe, NM 87505-4835, or e-mail orders@hmp.com. SoftStore, Inc., is a High Mountain Press Company.

Wholesale, Including Overseas Distribution

High Mountain Press distributes OnWord Press books internationally. For terms call 1-800-4-ONWORD (1-800-466-9673) or 505-474-5130; fax to 505-474-5030; e-mail orders@hmp.com; or write to High Mountain Press, 2530 Camino Entrada, Santa Fe, NM 87505-4835, USA.

On the Internet: http://www.hmp.com

OnWord Press, 2530 Camino Entrada, Santa Fe, NM 87505-4835 USA